why vegans have smaller brains

AND HOW COWS REVERSE CLIMATE CHANGE

David Ellis, Alison Morgan & Anita Tagore

WILD CAT

First published in 2025 by
Wild Cat Publishing
In partnership with Whitefox

www.wearewhitefox.com

HB ISBN 9781917523042
PB ISBN 9781917523059

Also available as an eBook
ISBN 9781916797130

Please go to the following website to access the full set of references, notes and sources for this book: http://whyveganshavesmallerbrains.com

All photographs and illustrations in this book © David Ellis, Alison Morgan and Anita Tagore, with the exception of the following: p. 34 (left) Shutterstock/Patrick Rolands, (right) Shutterstock/Mironov Vladimir; p. 54, Shutterstock/Valentyn Volkov; p. 138, Shutterstock/New Africa; p. 139 (top left) Shutterstock/Yakov Oskanov, (top right) Shutterstock/Sarymsakov Andrey, (bottom left) Shutterstock/Aleksej Starostin, (bottom right) Shutterstock/stavklem; p.234, Paris Bannerman & Alison Morgan.

Designed and typeset by Tom Cabot/Ketchup Productions
Cover design by Francesco Bo Coppola
Project management by Whitefox
Website design by Paris Bannerman

CONTENTS

INTRODUCTION

'Let food be thy medicine.'

Hippocrates

WHAT IF WE COULD STAY HEALTHY UNTIL we drew our very last breath? What if we could carry on living in our sixties, seventies and beyond as we did in our twenties? What if we were to tell you that the answer to good health lay on our breakfast, lunch and dinner plates? And what if, in the process of becoming and staying healthy, we could also save the environment?

The quote above, 'Let food be thy medicine', is attributed to the ancient Greek physician Hippocrates. But how might we interpret the saying? Today, most of us have bought into the idea that a plant-based diet is the best 'medicine', not only for our own health but also for the health of the planet. This book questions this doctrine. In fact, it goes so far as to advocate the complete opposite as prescription: an animal food-based diet. It sets out to challenge what we consider to be unquestionable truths about the benefits of veganism and vegetarianism, and to explain how these beliefs can harm our health, cost us money and damage the environment.

Why Vegans Have Smaller Brains, the blunt title for this book, has a simple justification. That's what brain scans show.

What has surprised us the most in our nutrition research is that breastfed babies of vegan or vegetarian mothers can develop brain atrophy. It is a lack of certain nutrients – nutrients that are not available in any plant food – that gives rise to this brain atrophy, or in other words, brain shrinkage. And studies at the opposite end of the age spectrum have found that a lack of the same key nutri-ents speeds up brain shrinkage as we get older. Importantly, these nutrients are found in ample amounts in animal foods. From the cradle to the grave, cutting out animal foods from one's diet has serious and debilitating consequences. The title of this book contains no exaggeration.

Vegetarianism and veganism, it is important to point out, can be choices that we make on ethical grounds, and this perspective is one with which we have sympathy. We feel no personal antipathy towards – nor do we wish to vilify – either vegans or vegetarians. We have been on the other side of the fence. One author was vegetarian in her teens and twenties, and, much to her chagrin, her daughter has followed suit for the past eleven years. This author was brought up in a home where meat was considered unclean and the cow sacred. She has therefore arrived at the conclusion that an animal-based diet is the healthiest diet not through a natural preference but despite it. And what she has learned has changed her life.

As a species, humans are living longer than ever, but not necessarily living better. Life expectancy has risen considerably in the past century. But at what price? Diabetes, dementia, diapers or being bedbound with a broken hip awaits many of us at the end. People in developed countries frequently spend their last decade suffering illness after illness, culminating in hospitalization and heavy medication in their final days. And, as lifespans increase in poorer countries, billions more people are heading in the same direction.

It is universally acknowledged that there is a link between diet and disease. If we make the right choices about what to eat, we'll be rewarded with longer, healthier lives. But this undisputed fact is only a starting point, and it leaves important questions unanswered: which foods, and why?

Ironically, given the power of modern science, deciding what to eat might be harder now than it has ever been. Ever since newspapers began to tell us about nutrition studies, we have been worrying about what to gobble down with gusto and what to avoid – but trying to make sense of the studies can leave us none the wiser. Some studies show that the so-called 'Mediterranean diet' is healthy; others indicate that a vegan diet is the one to go for; and there are even a few claiming that a meat-only diet deserves that stamp. In the 1970s, we were told that consuming eggs was bad for cholesterol and health, and then a few years later we were told that eggs didn't affect our cholesterol and that they may, in fact, be good for us. The history of margarine is another example. It was first adopted as a cheap alternative to butter, then promoted as a health food, later castigated because it was riddled with trans fats and more recently reformulated without trans fats.

The net effect of all this contradictory information is widespread confusion. It is almost impossible for most of us to know what to believe about what constitutes a healthy diet. Given the bewildering barrage of competing claims, contradictory evidence and scientific jargon, many of us have simply defaulted to what has become a widely accepted view: eat less food from animals and more plant foods instead. Today's diet heroes are vegetables, grains and fibre. Meat has been cast in the role of villain.

This confusion is owing to the highly questionable validity and soundness of nutrition studies. The discrepancies from nutrition research are due, in part, to the biases of the statisticians interpreting the data as well as the flawed designs of their studies. The well-known adage, 'There are three types of lies: lies, damned lies, and statistics', is epitomized in modern nutrition research.

One of the greatest failures of Western medicine is that it does not appreciate that most chronic illnesses are preventable and reversible with dietary changes. The main aim of this book is to address this failure by setting out a diet that prevents, treats and reverses common 'Western diseases' such as cardiovascular disease, diabetes and certain cancers.

We will use the Sapiens Diet to describe the animal-based foods that we argue are best for our species, *Homo sapiens.*

Along the way, we'll serve up a menu of topics that will inform us about what to eat and why. We'll explore why animal fat has been wrongly demonized for the past fifty years even though the evidence for this view was unclear from the outset, and also why the adoption of a low-fat diet has coincided with – and may have caused – an unprecedented rise in obesity and diabetes over that time. We'll also examine the many inconsistencies in nutrition, and, most importantly of all, we'll investigate fundamental misconceptions about animal foods and what exactly the Sapiens diet is.

The word 'meat' conjures up the image of steak or ribs or chops. We think of meat as muscle – the thing that tenses up on demand and helps us to move – but animal foods are so much more than just meat. Together, meat, fat and offal provide all the vitamins, minerals and calories that we need, including, to most people's surprise, vitamin C. Moreover, offal, the food that contributed to the evolution of the human brain more than any other, gives the most bang for your buck, and has more vitamins and minerals per penny than any other food. In terms of both value for money and total nutritional value, neither blueberries nor broccoli deserve to be considered 'superfoods'. That title fits another diet group far better: animal foods.

We'll also learn why the Sapiens diet reverses 'leaky gut syndrome' (and what on earth leaky gut is); why a lack of fibre does not cause constipation; how it is possible to get more than enough vitamin C on a meat-only diet and why animal fats are healthier than plant-based fats. We'll also find out why meat, when it is part of the Sapiens diet, does not increase the risk of heart attacks, cancer or gout.

But what use would it be if a diet that benefitted our health exacerbated global warming? For that reason, the focus of the third part of this book is to reconcile the Sapiens diet with the environment, and to show how it is possible to farm livestock in ways that benefit the ecosystem and also contribute to the fight against climate change.

Not many of us know that, after the oceans, the soil is the largest carbon store in the world. Soil contains more carbon than the atmosphere and all plant and animal life combined. Many farming methods degrade the soil and cause carbon to be emitted as CO_2 into the atmosphere. Around a third of the CO_2 that humans have added to the atmosphere has come from the soil as a result of farming over the last 10,000 years. But what many do not realize is that it is the growing of crops, such as grains and vegetables, that is the main offender. Not only does crop farming deplete the soil, but it also does untold damage to the ecosystem. Grassland and well-managed grazing livestock, such as cattle and sheep, do the opposite.

Contrary to what we are led to believe, reducing the number of farmed animals will not solve the climate crisis. After all, if we are saying that farming cattle is bad for the environment, then are we also saying that wild ruminants – such as the vast herds of wildebeest in Africa – are equally damaging to the planet? This question suggests that it is the way that animals are managed that is the important issue – it's 'not the cow, but the how'.

There is hope: as we shall see, it is possible to rear and manage animals in ways that benefit the soil, the environment, the nutrient content of food and the health of humankind. And one chapter will explain why there is no such thing as a truly vegan diet; how farming for animal foods is complicit in *fewer* animal deaths than farming for a plant-based diet.

A higher proportion of millennials and Gen Z are choosing to be vegan or vegetarian compared to the generations that have gone before them, and older people are also joining this band. There has never been a better time to talk about this important matter.

Human beings evolved because of the astounding nutrient content of the animals that they hunted or scavenged. Our bodies grew in tandem with the availability of animal foods that were packed with vital nutrients. The adoption of agriculture 10,000 years ago paved the way for civilization. However, the cost of civilization was a *deterioration* in our health. (It's a popular misconception that life, surely, 'got better'. No, it didn't. For example, the move to living in towns and cities led to diseases associated with poor sanitation; it's only recently that we've achieved clean drinking water and effective sewage systems. Some might say we still struggle with the latter.)

But, now that we are 'civilized', we can have the best of both worlds. We can go back to the diet that we evolved with, and which best suits our biology, as well as reaping the benefits of civilization.

Today, in spite of our evolution, meat has become a scapegoat for both our health problems and global warming. This book challenges current ideas, presents new findings from the world of nutrition, agricultural and environmental science, and offers practical solutions. And it acknowledges the link between what we eat with how food is farmed, because too often the interrelationship between the two is overlooked.

It's time for a major rethink about diet, health, farming and the environment.

Section One

History: Personal, Ancient, Modern

CHAPTER 1

HOW I ARRIVED AT THE SAPIENS DIET

Personal story of one of the authors

'The truth is rarely pure and never simple.'
Oscar Wilde

ONE FINE DAY IN HACKNEY, MRS TAYLOR walked into my consulting room for her ten o'clock appointment and proceeded to change my life.

Like most doctors, I was running late and started with my standard opening line of, 'Sorry to have kept you waiting,' before moving on to, 'What can I do for you today?' Slim and smiling, and with a bit of a twinkle in her eye, she replied, 'I received a letter asking me to come and see my GP to discuss my blood results.' Mrs Taylor looked fabulous and much younger than the sixty-four years her medical records showed her to be. She also appeared hale and hearty, and not obviously in need of any medical services. In fact, Mrs Taylor was the very picture of health.

After a quick look through her blood tests, I said, 'Ah, I see that your cholesterol level is much higher than it should be.' I carried out a few basic checks, such as her weight and blood pressure, which were normal, and then embarked on the lifestyle lecture that encouraged a low-fat diet and regular exercise. Normally a weight-loss diet is part of this spiel, but it seemed quite unnecessary here because her weight was already ideal. Mrs Taylor's cholesterol level, however, was extremely high – this was the reason why she had been asked to make an appointment to see one of the doctors. My advice to her was that she start some medication as well as making lifestyle changes. Mrs Taylor listened patiently before replying that she would rather not take any pills and that she would bring in something for me to read that would explain why she was declining the treatment. I had never questioned the thinking behind the benefit of lowering a high cholesterol level – I had always taken it as a given that the lower the level the better.

One week later, Mrs Taylor turned up without an appointment and apologetically popped into my room in between patients to drop off a book. Without any fanfare whatsoever, *The Diet Delusion* landed on my desk. I had expected something to do with pills, but instead, here was a diet book. Nonetheless, ever-curious, I started reading it as soon as I got home. It's not your typical page-turner, but I found it unputdownable. Six hundred dense pages later, I was a complete low-carb convert – and strongly against low-fat diets.

It was lucky that Mrs Taylor came to see me rather than another GP in the practice. As a teenager, I devoured diet and recipe books. I had developed a long-standing and passionate interest in food, nutrition and diets, and was open to new ideas. But why exactly was *The Diet Delusion* such a game changer for a forty-something GP? Authored by Gary Taubes, a Harvard-educated physicist turned science journalist, the book explained in detail why most nutrition studies are seriously flawed. And so, because nutrition guidelines are largely based on these studies, it follows that the guidelines we are given about what to eat are flawed too. I, like most, had been seduced into accepting the low-fat mantra that has been touted by mainstream nutrition science for the past fifty years. But there is, in fact, an ever-growing body of evidence that clearly shows that current nutrition guidelines are unreliable, and turns on its head what most of us have accepted for so long.

Reading the book, though, put me in a tricky position: I found it increasingly difficult to advise my patients to eat a low-fat diet. In fact, being forced to give what I now viewed as incorrect nutrition advice was one of the factors that motivated me to stop being a GP a few years later. My caution was not misplaced: other doctors, such as the orthopaedic surgeon Gary Fettke, have been censured by medical boards for giving low-carbohydrate dietary advice and have even been ordered to stop speaking about diets with their patients. And Tim Noakes, a professor of sports science, was tried for misconduct for giving low-carbohydrate dietary advice – this professor's fate hung in the balance for a full two years before he was exonerated.

Coming upon *The Diet Delusion* was one of the most serendipitous moments of my life. It rekindled my passion for nutrition. After quitting as a GP, I went back to studying and completed a master's degree in the subject. The book had come at a time when I wasn't as rushed off my feet as I had been when I was a junior hospital doctor or when my children were toddlers. Mrs Taylor wasn't just the trigger for my Damascene conversion to a new way of eating: she set me on a completely new career path.

Along the way I have learned that food is not only life-changing but also IQ-changing. For example, a mother's diet has effects on her child's IQ, as is the case in dietary iodine deficiency, which affects an astonishing 2 billion people. If the soil contains very little iodine, the food that grows out of it will contain very little iodine too – such as in the foothills of Mount Everest. If levels of iodine in the food of a pregnant mother are low, her baby's brain development can be affected and the baby has an increased risk of being born with a condition called congenital hypothyroidism, which results in severe and permanent intellectual disability. In the 1950s, the incidence of this condition in the Sherpa people of this remote region used to be as high as one in eight. This story luckily has a happy ending. In the 1960s and 1970s, potential mothers were given iodine, after which this type of intellectual disability disappeared completely in the next generation. It has been estimated that eradication of some of the world's most common nutrition deficiencies – iodine, iron and zinc – would shift the world's IQ up by ten points.

As if being IQ-changing weren't enough, food also affects our *genes*. Genes are our basic units of heredity, passing on physical characteristics and traits from one generation to the next. Even though genes themselves can't be changed after they have been inherited at conception, the *effect* of a gene can be changed. This is because genes can be switched on or off by various external and internal factors, including diet. The idea that there is an interaction between genes and the external environment isn't revolutionary: what's new is the discovery of *how* environmental factors cause genes to be switched on or off. It turns out that your genes are not necessarily your destiny.

The way that genes can be turned on or off by environmental factors is being explored by researchers in the new and exciting field of 'epigenetics', which aims to explain how nature and nurture are linked. An easy to understand example of how epigenetics reveals what effects our physical characteristics is height. It turns out that only 80% of an individual's height is determined by inherited genes – which leaves 20% under the influence of other factors such as nutrition and culture. Studies on immigrant families show that moving to a country with better access to nutritious food is associated with increased height in the next generation. Genes do matter, but clearly environmental influences matter too.

Some of the external environmental factors that influence our genes include drugs, chemicals, temperature and light, and some of the internal factors include hormones and metabolism. If someone is found to have high blood pressure, high cholesterol, diabetes or gout, they may end up with a lifelong prescription for tablets to treat the condition. Just as Mrs Taylor declined drug treatment for her high cholesterol, many think twice before embarking on such medication – the idea of taking drugs till the end of one's days can be daunting. However, what we eat also goes into our body and permeates it just as a drug does. Food, therefore, is also an environmental factor that influences our genes. But do we take decisions about what to eat as seriously as which drugs to take? Do we treat them as equally important? No, we shouldn't. Food is far more important. Food is something that our bodies encounter every single day throughout our lives, and their effects add up over the course of our lives.

The effects of food on our genes starts even before we are born: the food that a mother eats when she is pregnant has an impact on her child's risk of developing diseases such as diabetes, heart attack or stroke in adulthood. The good news is that, if your genetics predispose you to a disease, epigenetics can offset some of the risk.

Epigenetics can be very revealing about the causes of good health, and one of the best examples is what we know about how the diet of pre-pubescent boys can affect their grandson's health. Dr Lars Bygren, a preventative health expert, studied a group of men born in 1905 in the north of Sweden to see if there were any long-term effects of the feast and famine years of the previous century in that area. What this research showed was that a father's experiences in childhood have effects on his sperm that are then transferred to his male descendants, and a mother's experiences as a foetus and infant are transferred to her female descendants.

It used to be thought that the genes we are born with cannot be changed. We now know that this blueprint is not set in stone. Yes, there are cutting-edge gene-editing techniques, but when it comes to everyday life we ourselves can change what effects our genes have on us and on our descendants, with our diet. Some say that we are what we eat: perhaps we should add to that adage 'we are what our parents and our grandparents ate'.

Personally, I would feel vindicated if I could put some of the blame for my propensity to diabetes and being overweight on my ancestors, especially my tyrannical grandmother. I grew up in the 1960s in a Hindu family in India where three generations, from my grandmother down to me, lived together. Luckily, my parents decided to move to England when I turned eleven and I escaped the oppression of living in an extended family. This transcontinental move happened to coincide with my going through puberty. The changes that puberty induced in me were almost like sorcery: I went from an awkward and chubby child into an awkward but slim and svelte adolescent. However, not only did my shape and size change, but my skin tone became lighter in colour too – a gladly received bonus for a dark-skinned Indian girl. Was it the change in my diet, from chapatis and lentils to roast beef and Yorkshire pudding, or was it the dramatic changes in my hormones that induced these changes? Even though both food and hormones have extraordinary effects on our bodies, in puberty, hormones win.

As a teenager I was vaguely aware of the fact that all the changes in my body must have had something to do with my hormones, but I wasn't too concerned with the why or how. I was just every grateful for the beneficial transformation. Hormones, however, are of the utmost importance. For example, the hormone oxytocin is responsible for contractions during labour and the production of milk during breastfeeding. However, oxytocin has another side too: it is released during sex and is partly responsible for the intoxicating feeling of falling in love. Some have aptly dubbed oxytocin 'nature's ecstasy'. Touching, kissing and sex all flood your brain and body with oxytocin, and oxytocin is one of the few hormones with a positive feedback loop: once it starts going up, its production goes up exponentially – which is why we get that rollercoaster-ride feeling when falling in love. It is no wonder that many 'friends with benefits' end up becoming emotionally attached. We can blame our hormones – they can make it difficult for us to distinguish lust from love.

There are countless other examples of the obvious and not-so-obvious effects of hormones. One not so well-known effect is the fullness we feel after a meal, which is partially down to a hormone called leptin. A more obvious example is what happens during puberty: in girls, oestrogen levels go up, which causes a narrowing of the waist, extra fat to be laid down in breasts and on hips, and a lightening of the skin; boys, on the other hand, develop testosterone-driven bulging muscles and less fat on their hips. Hormones determine *where* and *how much* fat, muscle and bone will grow. During their growth spurt, teenagers seem to be always hungry and massively increase their food intake. But is it their overeating that causes their growth spurt? Clearly not. Rather it is the growth spurt that makes them hungry. And once they are out of this phase, they won't grow any taller, no matter how much they eat. Hormones win again.

Of relevance to this book are the dramatic effects that different foods have on hormones. Did you know that fat, protein and carbohydrates all have very different effects on hormones? If we were to compare the effects of 500 calories from a lamb chop to 500 calories from a slice of cake on our hormones, we would notice stark differences. A more extreme example would be to compare eating 500 calories of pure lard with 500 calories of sugar. As they say, 'a calorie is always a calorie' but because our bodies process different foods in different ways, the effects of calories from fat, protein and carbohydrates are not necessarily the same. Even though it sounds counter-intuitive, eating pure sugar has far worse effects on your hormones than eating pure fat would (although both these scenarios might make you feel a bit nauseous).

Why counter-intuitive? Because it sounds logical that eating fat will make you fat. After all, scientists, dieticians and the media have long promoted this idea: that is how the widely held misconception that 'a low-fat diet is healthy' has gained traction and credibility. When it comes to food, one of the more important hormones is insulin. How much fat you gain after a meal depends entirely on how much insulin is released: the higher the insulin level, the easier it is to convert food to fat. Bizarre though it may sound, a meal of pure fat, without any carbohydrates or protein, will *not* cause a significant release of insulin and therefore will not result in accumulation of body fat.

We have all bought into the idea that eating fat will make you fat. But this isn't the case; it is eating carbohydrates that makes you fat. How do they do this? Carbohydrates make insulin shoot up – not meat, not animal fat, but carb-loaded food – bread, cereals, rice, pasta, potatoes, fruit, beans – all the types of food that most dieticians insist should replace meat and animal fats. What I have learned over the years is that the very same foods that we're being told to eat are exactly the ones that are undermining our attempts to fight off obesity.

There is yet more to insulin: besides governing how fat you are, high insulin levels also lie at the bottom of many diseases – heart attacks, strokes, dementia, diabetes, certain cancers and even skin conditions such as acne. Owing to its importance, insulin will be covered in more detail in chapter 5.

How much insulin is secreted after eating carbohydrates appears to be genetically determined, which is why some people find it easier to stay slim than others. I, sadly, drew the short straw in the weight-control gene stakes. Try as I might, it was a constant struggle to maintain my new slimmed-down adolescent physique. I believed, as much of the world did, that one's weight was determined simply by a balance between calorie intake and exercise. So, after the hormone-induced metamorphosis of my own body, I took up regular exercise and went from being a couch potato in the library to jogging four miles before breakfast and playing squash several times a week.

But the 'eat less and move more' advice didn't work for me. Little did I know then that it doesn't work for most people. I should have realized that it wasn't exercise that brought about the pubertal changes in my body in the first place, and so exercise was unlikely to help me keep my new figure. It was all about hormones, which most of us assume we can't do much about.

When it dawned on the teenage me that exercise wasn't working, I started to look at what I was eating. Thus, at the tender age of thirteen, my lifelong interest in diet began, and I was hurled onto the path of becoming a serial dieter. First stop was the Atkins Diet (a low-carb diet). Next stop was vegetarianism (one of the few big regrets in my life). In my early twenties, I combined vegetarianism with the recommended low-fat diet. It doesn't get much healthier than this. I also took up fasting two days a week (as all good Hindu women used to) and routinely used to skip breakfast (despite being repeatedly told that one should 'breakfast like a king'). None of this, though, was enough to stop the pounds from piling on.

Like most people, I was brainwashed into believing that a calorie-controlled diet containing little or no red meat or animal fat but with more olive oil, fruit, vegetables and whole grains, washed down with a glass of red wine, was the healthiest diet. This 'Mediterranean diet' was originally based on a close-knit rural community on the Greek island of Crete working outdoors in the sunshine in the 1950s. These happy Cretans ate very little in the way of sugar or ultra-processed foods and they fasted for up to 180 days a year. Yet, rather than assign their longevity to happiness, sunshine, fasting, a lack of sugar in their diets or good social networks, dieticians put their long lives down to olive oil, whole grains, pulses and Pinot Noir.

Convinced, I decided to try this supposedly healthy Mediterranean diet. I diligently counted calories, ate whole grains and suffered the switch to olive oil (I might be the only middle-class person who abhors the taste of olive oil). I shunned red meat completely and brought up my three children as pescatarians (an even bigger regret). And, while other new mothers enjoyed a well-deserved rest while their young children were asleep, I would be pedalling furiously on a stationary exercise bike in my living room. Despite a calorie-controlled Mediterranean diet and an insane exercise regime, my waistline slowly expanded. Worse yet, minor skin issues that I had suffered in my teens came back with a vengeance: eczema, vague rashes, acanthosis nigricans, pre-menstrual acne, angular stomatitis, dermographism, cold urticaria and allergies to perfumes and make-up. It was the unhealthiest time of my life.

Then, in 2012, by chance I came across another ground-breaking book, *Food and Western Disease*. The book's author was Staffan Lindeberg, a Swedish professor of medicine. It was based on his doctoral thesis about the diet and health of a population living on Kitava, a remote tropical island in Papua New Guinea in 1989. This was one of the few remaining populations on earth with dietary habits matching what would have been the case for *Homo sapiens* in pre-agricultural times.

At that time, the dietary staples of the Kitava people were fish, meat, fruit and vegetables. As far as their health went, what Professor Lindeberg found was that the weight trajectories of the Kitava population moved in the opposite direction to those of most Westerners – they became slimmer as they aged, despite an abundance of food. Furthermore, unlike those who consume typical Western diets, the insulin levels of the Kitavans also went down with age. The mean insulin level in middle-aged Kitavans was an astounding 50% lower than that of middle-aged Swedish people. These markedly lower levels of insulin might be one of the reasons why the Kitavans lived long and healthy lives, and why heart attacks and strokes were virtually non-existent – even though 78% of the population were smokers

Coming back to my personal story, another rather peculiar change that I found in myself after taking up a Palaeolithic diet nine years ago was that some of my grey hairs started going back to black. Similarly, the abnormal skin pigmentation (acanthosis nigricans) that I used to have, and which was very pronounced on my face, elbows, knees and toes (sounds like the song!), improved when I took up a Palaeolithic diet. I then discovered that the skin pigmentation of two of my friends also improved with this diet.

It doesn't sound like a life-and-death matter to have abnormal skin pigmentation, but acanthosis nigricans is strongly associated with abnormally high insulin levels and often precedes full-blown diabetes. It is, therefore, potentially very serious. Interestingly, insulin levels are known to get worse in pregnancy and, when I was pregnant with my eldest child thirty years ago, my acanthosis nigricans flared up so badly that I ended up consulting my GP about it. However, at the time, the association between acanthosis nigricans and abnormal insulin levels wasn't known. Neither my GP nor I could fathom why the skin on my neck had suddenly become thick and almost black when I was pregnant (it used to be thought that acanthosis nigricans was mainly associated with certain cancers). In retrospect, I can also see that I looked permanently tired, with pronounced dark circles around my eyes (also owing to abnormal insulin levels).

A friend who was diabetic, overweight and had high blood pressure, kidney problems and acanthosis nigricans took up a Palaeolithic diet on my recommendation. After two years of following the diet 90% of the time, her weight came down and there was a big improvement in the pigmentation on her face. The pigmentation on this friend's knees and elbows had all but disappeared too. These acanthosis nigricans blemishes are not uncommon on the skin of Indians (who also happen to be at a higher risk of diabetes than white people are).

Furthermore, this friend's blood pressure and blood glucose came down to such an extent that she could come off her diabetes and blood pressure medication. Her abnormal kidney function, which is par for the course in those who suffer long-standing diabetes, also normalized after a few years of changing her diet. In all my years of working as a GP, I had never seen the abnormal kidney function of a diabetic patient being reversed. Ten years on and my friend's blood pressure, kidney function and blood glucose are all still normal. With only a simple change in diet, she has managed to stay off medication, which in turn has produced a massive saving in drug costs for the NHS.

As far as my own conditions go, angular stomatitis, which is due to an iron and B vitamins deficiency and shows up as cracks at the side of one's mouth, has resolved completely. The most dramatic of my skin conditions, though, was dermographism, which literally means writing on the skin: a mere scratch would generate a florid red weal and you could, in fact, write your name on my skin by lightly scoring it. This condition sometimes resurfaces after just a single cheat meal.

The other big influence in my diet-research journey was stumbling upon the fact that epilepsy can be successfully treated by a particular type of diet: a ketogenic diet (popularly known as the 'keto' diet). In short, a ketogenic diet consists of large amounts of fat, moderate amounts of protein and very little in the way of carbohydrates. This way of eating forces your body to start using fat as a source of energy, which it does by converting fat into 'ketones'. On a standard Western diet, ketones are not routinely produced, and glucose is the main fuel that the body uses. On ketogenic diets, however, our bodies produce and use both glucose *and* ketones.

What made me sit up and take ketogenic diets seriously was when I read that children with epilepsy could be cured by following a ketogenic diet. Some children who were having twenty fits a day, despite taking multiple anti-epilepsy medications, stopped having fits and could come off and stay off all their medication. But only as long as they kept to the diet.

Serious bowel diseases, such as Crohn's disease and ulcerative colitis, have also been successfully treated with a ketogenic diet. I read about patients who went from suffering with bloody diarrhoea twenty times a day for many years to normal bowel movements in the space of a few weeks after adopting a ketogenic diet.

Besides treating epilepsy and these serious bowel diseases, ketogenic diets also help overweight people lose weight. The benefits of being in long-term ketosis are, in fact, manifold: a better-functioning immune system, reduced inflammation, protecting against infections and cancer, and beneficial epigenetic effects too.

Despite the enormous promise of a ketogenic diet, I still wondered about whether it might be dangerous or unhealthy. Understandably, you might be wondering the same thing too. Eating lots of fat and meat, and not much in the way of whole grains, fruit or vegetables is pretty much the opposite of what is commonly regarded as a wholesome or sustainable diet. But we will cover in detail throughout this book why eating in the opposite way to what is currently recommended is a much healthier and more sustainable way of eating.

Besides treating epilepsy, ketogenic diets show potential in treating other disorders such as Parkinson's disease, Alzheimer's disease and brain tumours. Surprisingly, ketogenic diets also show promise in treating erectile dysfunction, psychiatric conditions and mood disorders. Ketogenic diets could be the new Valium and Viagra rolled into one!

An issue that I, like most doctors, used to have with ketogenic diets was that they were 'unnatural' and unsafe. But I have had to let go of my original, ill-informed suspicion of ketogenic diets – we will learn in chapter 11 why ketogenic diets are safe and actually more natural than current recommended diets.

Success in the treatment of all these conditions with a ketogenic diet, though, is not 100% guaranteed. Moreover, ketogenic diets are sometimes associated with nutrient deficiencies and stunting in children. But ketogenic diets come in many different shapes and sizes. It turns out that there is one particular type of ketogenic diet that is *not* associated with adverse side effects: the Palaeolithic ketogenic diet (termed the 'Sapiens diet' in this book).

What exactly does this magic diet look like? The Sapiens diet is a type of ketogenic diet and consists only of foods that would have been eaten during our evolution. It consists of large amounts of animal fat, moderate amounts of offal, meat, fish and eggs, and optional small amounts of vegetables and fruit and very small amounts of honey. A Sapiens diet differs from other ketogenic diets in that vegetable oils, milk products and nuts are not allowed. Furthermore, it is the only diet, ketogenic or otherwise, that can supply *all* the nutrients that we need – every single mineral, vitamin, fat and protein, including vitamin C and vitamin D. In fact, the Sapiens diet corrects rather than causes nutritional deficiencies. There is absolutely no need for any kind of supplement – no extra vitamin D tablets or probiotics or anything else.

You may also be thinking that the diet we are advocating doesn't sound very exciting. Don't we eat for pleasure? (Although we should remember that's a privilege that not everyone can afford.) What about the delights of making and breaking bread, hosting dinner parties, going out to cafes and restaurants, poring over glossy, celebratory-chef recipe books of spicy and exotic dishes, and so on? Wouldn't Sapiens deny us those joys? Well, we do a lot of things for pleasure. Many used to smoke. Many drink alcohol. But pleasure doesn't equal health; we know the damage that smoking or over-drinking can do. 'What many don't know is that much of our modern diet, while enjoyable, may be consigning us to the diseases of the Western world, including a decline into decrepitude in our old age, or earlier if we develop type 2 diabetes.' And besides, Sapiens meals can be delicious (think slow-cooked pork belly or beef brisket, served alongside a few roast carrots).

Usually, when something sounds too good to be true, it *is* too good to be true. So, I went in person to the Hungarian clinic where these extraordinary claims originated, and I met the research team to see their results first-hand. What I learned was that the findings from the clinic were reliable. The team have seen and treated upwards of 10,000 patients. Virtually all the patients who keep to the diet manage to reverse their medical conditions. When patients lapse away from the diet, their conditions return. And when they go back on the diet again, they get better again. That is much better validation than the average diet study.

Initially, the clinic used a diet of meat, fat and offal to treat patients with Crohn's disease and went on to include it as a treatment for another serious bowel disorder known as ulcerative colitis. When the team running the clinic found that the diet had other effects, such as the reversal of diabetes, obesity, high blood pressure and other autoimmune conditions like rheumatoid arthritis, they started recommending the diet as a standalone treatment for all those other conditions as well.

While researching this book, it made sense to also investigate how the human diet evolved. Besides discovering that diet was crucial in the evolution of the human brain, it struck me that, when asked about the big cats (be it mountain lions in America, Bengal tigers in India or cheetahs in Africa), no one would dispute that they are all carnivorous animals that thrive best on raw meat. Nor would anyone suggest that these animals might benefit from the addition of a bit of lettuce or other greens to their diet. And the idea of putting lions on a diet based on their blood type sounds ridiculous. Why then do the likes of 'the maple syrup diet', 'the blood type diet' and other hokum diets appeal to humans? Just as there is a diet that each species of animal evolved with and which still suits them best, so it is with humans too. The colour of your skin, the temperature of your environment and such like are irrelevant when it comes to diet. No matter where we live or what our ethnic origin might be, genetically we are all *Homo sapiens*. Just as bison on the steppe of Asia and on the American prairies need to eat the same diet, so too should humans from the Arctic to Zambia. If we look at the human diet from an evolutionary perspective, we can see that the

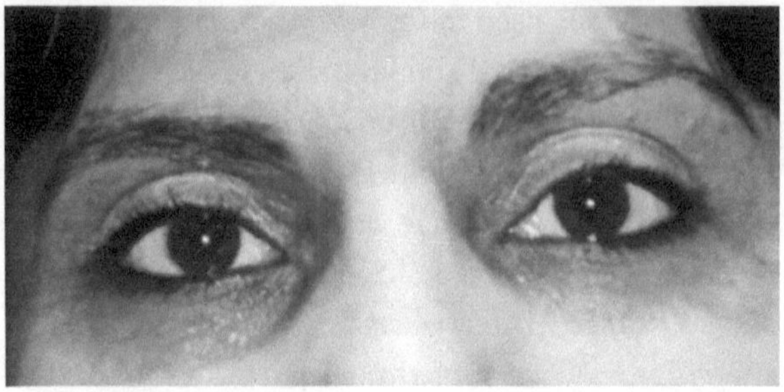

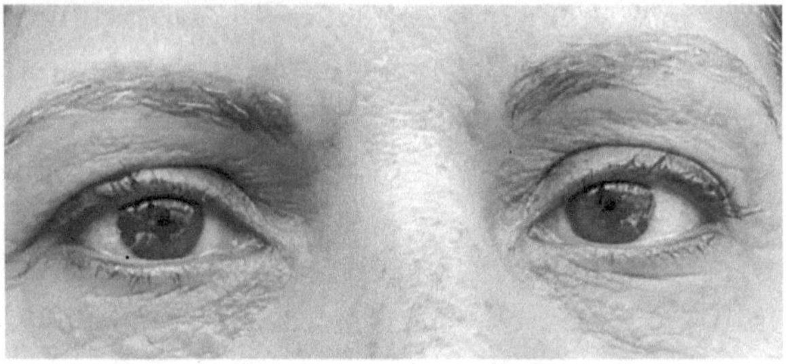

Pronounced pigmentation around my eyes in February 2012 and recent photo (March 2023) with no abnormal pigmentation around my eyes.

Sapiens diet is the ideal diet for *everyone*. The evolutionary rationale for the diet is explored in detail in the next chapter.

After coming across the Sapiens diet, the next step was to try the diet myself. To my amazement, following the Sapiens diet for only four weeks caused a complete reversal of my leaky gut, which I previously didn't even know I had. (Leaky gut is such an important issue that it gets its own chapter later.) What I had been following before the Sapiens diet was a Palaeolithic diet that consisted of lots of fruit, vegetables and a moderate amount of meat, fish, eggs and nuts. It sounds fairly healthy, doesn't it? I then did the opposite of what most nutritionists would have suggested – I continued avoiding grains, *reduced* my fruit and vegetable intake by about 90%, cut out nuts and massively *increased* the amount of animal fat in my diet. One month later, in addition to correcting my leaky gut, my weight also went down. I continued following the diet most of the time over the next year and, besides seeing further improvements in my leaky gut, the reversal of the greying of my hairs accelerated and the pigmentation of the skin around my eyes disappeared completely.

Naturally, the next stop for me was to start testing out the Sapiens diet on my nearest and dearest. The first friend who was willing to take up this gauntlet agreed on the condition that he would follow the diet for only two weeks. I was sceptical about whether there would be any changes in such a short period of time, but his results were even more dramatic than mine. And after a couple of months of being back on his usual diet, his leaky gut test became abnormal again.

I was lucky that I was only suffering from minor skin conditions. Others, such as Mikhaila Peterson (a lifestyle blogger with over a million followers), have lived with severe diseases such as juvenile rheumatoid arthritis, which affected her so badly that she had to undergo many joint replacements as a child, as well as strong immunosuppressive treatments. Mikhaila found that eschewing all food except for meat remedied all her ailments.

But food not only affects physical diseases: it also appears to be a factor in severe mental health conditions. Dr Georgia Ede, a Harvard-trained psychiatrist, has found that low-carb and ketogenic diets have helped her patients with their psychiatric disorders, and many of them manage to come off all medication.

A diet of pizza, doughnuts, beer and fizzy drinks will not cause changes overnight, just as smoking one cigarette a day for a few days won't. It is what you do every day that matters. Regularly eating large amounts of grains, sugars, vegetable oils and plant foods can cause a leaky gut, and a lifetime of excessive carbohydrate and alcohol consumption will cause fat to spread insidiously around your liver and other organs. This is especially harmful to our health because this fat around our organs – known as 'visceral fat' – does more damage than fat in any other part of the body. Even if you don't eat to excess and don't appear to be overweight, you might yet have visceral fat or a leaky gut or both – being outwardly slim doesn't always mean that you are healthy. Most of us will know of someone who might have been slim and fit and yet succumbed to cancer, a heart attack or diabetes. Even young, ultra-slim and high-level athletes can be susceptible to diabetes. Take Sir Steve Redgrave, an Olympic gold medal winner, for example. In his prime – at the age of just thirty-five, and during a streak that saw him win five consecutive Olympic gold medals – Sir Steve was diagnosed with diabetes. Besides associating slimness with health, there is also a general feeling that being vegetarian will make you healthier. Sometimes I hear people say things like, 'Not only was she slim and fit, but she was also vegetarian. And she still got breast cancer.' Silently, I think to myself, 'Maybe being vegetarian was *why* she got cancer.'

The other surprising discovery that I came across after learning about the Sapiens diet is that offal provides all the vitamins and minerals that we need to live well. When I was young, one of the reasons why I decided to study medicine was the rather naive idea of being able to 'help people' by improving their health. What I have learned recently is that, if everybody in the world took up eating offal from regeneratively farmed animals, not only would health improve universally, but it would also benefit the environment, particularly in drought- and desert-stricken countries, which is where many of the poorest of the world live (we will explain what regenerative farming is in the final section of the book). If it were up to me, I would change the current dietary recommendations to include regeneratively farmed offal for all. After all, offal provides *more* minerals and vitamins (including vitamin C) than any fruit or vegetable does, and at a much lower cost – both to one's pocket and to the environment. The much-touted UK 'Eatwell Guide' should be relaunched as the 'Sapiens Guide'.

After learning about the pivotal role of offal and meat in the evolution of the human brain, I now eat a nutritionally well-balanced diet that consists mostly of animal fat, meat and offal. This diet was not the easiest diet to transition to, but I bit the bullet and went all out with it for a whole month in 2016, and intend to keep to the diet at least 80% of the time for the rest of my life.

This may not be perfect, but it is infinitely better than not doing it at all. I wish that I could say that I stick to the diet 100% of the time, but I have to admit to indulging in the occasional glass of wine, croissant or pizza. And every so often I do go overboard with fruit, vegetables or cheese. When I was young, bread was one of my staple foods and steaks were considered 'a cheat'. Now, the tables have turned, and it is bread that is the cheat, along with chocolate, in all of its many and wonderful forms. As far as diet goes, I have come full circle – from eating a low-fat vegetarian diet in my teens to a high-fat, meat-filled diet in my sixties. For me, there is no turning back. Despite not adhering to the diet perfectly, I have been consistently in ketosis for the past six years. These have been the healthiest years of my life.

In sum, what you eat is of paramount importance. We all need to be open to the idea that food can be both the cause of and the solution to all sorts of health problems. Proper nutrition offers one of the cheapest and most effective ways of improving health and reducing the risk of disease. Meat, animal fat and offal are the staples of the Sapiens diet. This book makes the case for these foods. It explains why they are the healthiest foods in the human diet, and why it is foods other than these that are the cause of ill-health. Our understanding of the potential of diet to change our health for the better is slowly growing and, with it, we can start to glimpse the possibilities. Food has the power to transform our health for the better – but only if we're ready to question our fundamental assumptions about what we should eat and why.

As well as food's impact on health, we also need to consider how it is farmed. This brings us to major flaws in the way people talk about diet, farming and the environment. Too often, dietary recommendations take no consideration as to how farming can deliver that diet, and the subsequent effect that changes to farming may have on the environment. And too often – in fact almost always – environmental researchers take no account of the way changes that they recommend for farming then impact our diet. Diet,

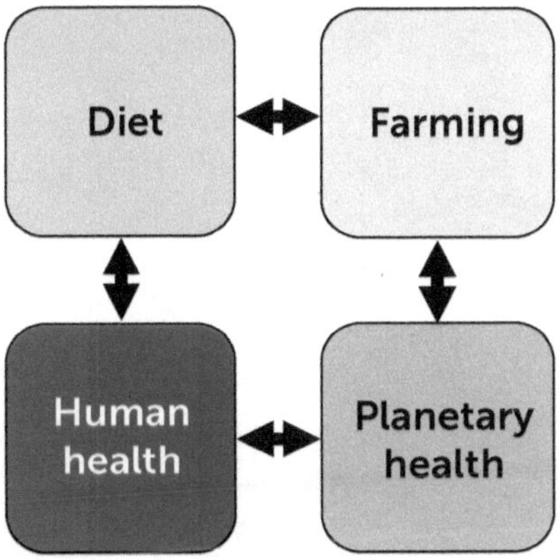

Our health, what we eat, how food is farmed and the environment are interconnected.

farming, environment; we can't change one without changing the others. What's needed is a holistic approach, an approach that acknowledges that as active participants on this planet everything we do has some influence on the environment. A dogma- and guilt-free approach that doesn't forget why we farm – to put food on the table, without which we cannot survive. Diet and the environment are irrevocably interrelated.

This integrated model is summarized in the diagram (left).

Health → Diet → Farming → Environment

A significant change in diet, perhaps in response to health issues, will require farming to react to supply that diet; this may then have a consequence for the environment depending on how that farming is carried out.

Environment → Farming → Diet → Health

And similarly in the other direction: changes in farming practice driven by environmental concerns may imply a change in the choice, cost and healthiness of foods offered to consumers.

Pity the farmer caught in the middle.

This book seeks to address all four elements in the model. Let's start at the very beginning – a very good place to start. The next chapter describes our ancestral diet; how we ate during our 6-million-year evolution to become 'thinking man and woman' – *Homo sapiens.*

CHAPTER 2

THE MAGIC OF MEAT

Humans didn't evolve to eat meat. Humans evolved *because* they ate meat.

'I read the prescription. It ran:
1 lb beefsteak ... every 6 hours.'
Jerome K. Jerome, *Three Men in a Boat*

AROUND 6 MILLION YEARS AGO, A NEW type of primate appeared that was different from its ancestors. It had a larger brain and walked somewhat upright. These early human-like primates – known as 'hominins' – were taking their first steps in our evolutionary journey.

Around 4 million years ago, the hominin *Australopithecus* arrived on the scene. Its evolution was marked by routinely engaging in a behaviour that its close relative, the chimpanzee, almost never showed – scavenging. *Australopithecus* began picking away at carcasses left at kill sites. The slow metamorphosis from ape to modern human began with a crucial change in diet.

Around 3 million years ago, hominins left the receding forests of Africa for the more arid savannah. At this time there was a global cooling and the air became stripped of moisture. Temperatures in Africa plummeted. Humid forests and woodland shrivelled away, leaving wide open spaces in their place. In this new, colder and drier environment, digestible plants were harder to come by. Animals that relied solely on the forests started dying out.

Fortunately for us, our hominin ancestors remained adaptable – when the world changed, they changed with it. They continued adjusting their diet and evolved. As luck would have it, the savannah was home to an abundance of grazing animals, and over time hominins became more reliant on hunting and the meat it provided. Having a 'head' start, with a brain that was already bigger than that of a chimpanzee, placed them well – a bigger brain meant a greater intellect. Out on the plains, they also became more skilful bipeds. Standing tall on their feet freed their hands to handle tools and weapons. From the need to adapt, and with the manual dexterity to exploit the newfound bounty on the plains of Africa, an upright animal-eating hunter was born: *Homo erectus*.

Then 300,000 years ago, during the final and most recent chapter of our evolutionary history, *Homo sapiens* appeared. The idea that humans evolved in an idyllic Garden of Eden, and that we were peaceful, plant-eating animals, is an appealing one. But this rose-tinted utopia couldn't be further from the truth. Over the last few million years, humanity has undergone a transformation from opportunistic scavenger to skilful hunter. Eating meat did not just coincide with that transition, it *caused* it.

HOW DIET SHAPED THE HUMAN BRAIN

Our brain is the quintessential feature that distinguishes us from the rest of the animal kingdom. If we compare the brains of our hominin ancestors with those of modern-day humans, we find that our brains have tripled in size over the last 4 million years. At the beginning of the evolutionary journey, our ancient forebears had a typical brain size of 440cm^3. That's about the size of a bread roll. But by 10,000 years ago, *Homo sapiens*' brain had grown three times as big, to 1,350cm^3 – roughly the size of a small sourdough loaf. Interestingly, in the last 10,000 years, our brain size has decreased a little, from 1,350cm^3 to 1,260cm^3. To complete the analogy, that's a bite-size profiterole's worth that has been lost.

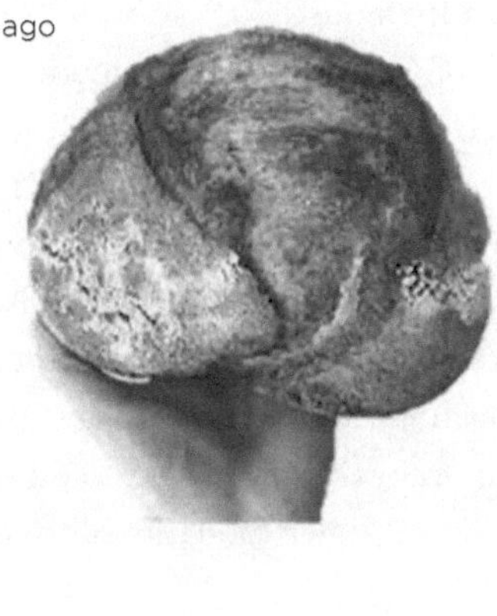

Use your loaf – evolution of hominin brain size.

Brain size over the past 10 million years.

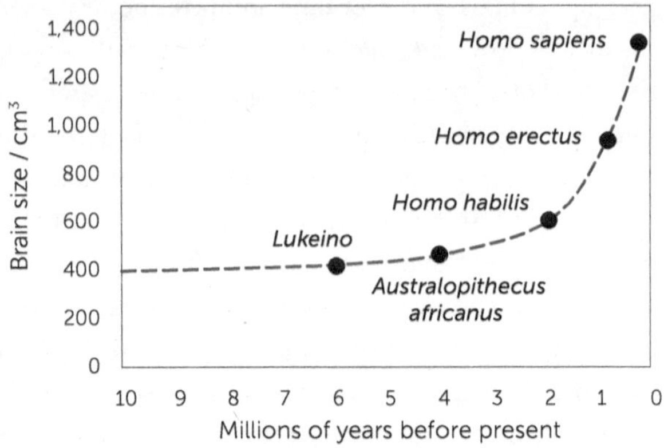

A large brain, though, is 'expensive'. No, there isn't a price on the human brain – what we mean by 'expensive' is that compared with other organs, the human brain greedily consumes a disproportionate number of calories. Furthermore, to grow a brain from the bread roll of *Australopithecus* to the loaf of *Homo sapiens*, the brain needed not only more fuel – in the form of calories – but also specific vitamins and certain types of fat. Importantly, only an animal-based diet could provide the extra calories *and* the crucial fats and vitamins, such as vitamin B12 and certain omega-3 fats. You couldn't pop in to Wholefoods to pick up your B12 and omega-3 supplements – we had to get them from our food. It was our ancient animal-based diet that kick-started brain expansion.

Evolutionary biologists say that high-quality (i.e. animal) foods were critical in the evolution of the human brain. Evolutionary biology is the branch of biology that is concerned with how evolution produced the diversity of life that we have today – it integrates information from anthropology, archaeology, biology, and many other 'ologies', to arrive at a theory of how we, and other species, evolved. They are right on the money with animal foods being high-quality foods.

Compared to plant foods, animal foods pack more nutrients and energy into smaller portion sizes. Meat, fish and eggs are calorie-rich and contain relatively large amounts of fat and protein, along with very generous helpings of vitamins and minerals. Plant foods may be richer in carbohydrates, but they are significantly poorer in vitamins, minerals and calories. Besides, carbohydrates are not even an essential nutrient for humans. Furthermore, plant foods are much more difficult to digest than animal foods. How can this be? Is it really easier for the human gut to digest a steak than a cucumber? Digestion involves breaking down food into substances that can be absorbed and used by the body;

a cucumber passes through the gut more quickly than a steak does, the gut therefore has less time to absorb the nutrients from a cucumber than those from a steak. Moreover, cucumbers contain fibre and anti-nutrients which hinder the absorption of nutrients (more on anti-nutrients in chapter 8). These are some of the features of plant foods that make them low-quality foods. Animal foods are, in fact, the epitome of a high-quality diet.

Learning that animal foods have more vitamins and minerals than plant foods has been a complete revelation to one of the authors. As was the discovery that it was the consumption of animal foods that got the brains and the intelligence of our ancestors going. Why on earth did she spend so many years trying to get her children to eat fruit and vegetables? To make matters worse, she denied them meat. She was, in fact, restricting the most nutritious of all foods. In her zeal to push plant foods on her children had she inadvertently put them at risk of nutrient deficiencies? Thankfully, they were allowed to eat fish, eggs and dairy foods.

Unfortunately, she was successful in making her children develop a taste for fruit and vegetables. She now has an uphill battle in convincing them to stop eating plant foods. Trying to convert a twenty-something vegetarian daughter to eat meat is a tough job. To date, she hasn't been successful.

PAYING FOR OUR LARGE BRAINS

As mentioned earlier, the human brain is an 'expensive' organ in terms of its energy usage. But how exactly do we pay for our gas-guzzling brains? It all started with eating meat. Over millions of years, eating meat led to a reconfiguration of our guts. How so? As the amount of animal foods in our diets went up, the human small intestine lengthened. At the same time, the colon shrunk considerably. This refashioning of the gut occurred because the small intestine is where nutrients from animal foods are metabolized and absorbed, and the colon is where fibrous plant foods are broken down. But besides this reshaping, there was also a net reduction in the overall size of the gut. The upshot of all this was a much lower energy usage by the gut. This energy windfall is what enabled our brains to grow larger. In effect, our guts and brains co-evolved.

In contrast, the chimpanzee gut has remained unchanged – it is still large and voluminous. Why does this matter? Because it tells us that the configuration of the human gut is more like that of a carnivore than that of a herbivore. Over the course of the last 6 million years, the growth of the human brain was balanced by an almost identical reduction in the size of the gut, which is now adapted to our preferred diet – high-quality, nutrient-rich, low-fibre, animal-based food. The long and the short of it is that the human small intestine became longer, and the large intestine became shorter.

THE BRAIN AND BONES BONUS

Seeing as 'meat' is the focus of this chapter, we will now take a closer look at the different types of meat. Most people think of meat as muscle – the fleshy material that moves an animal's body – or as the food equivalent: a chop, a loin or a fillet. This way of using the word 'meat', however, is limited. Animal food can also include organ meat, or 'offal', which includes brain, liver, kidney, marrow and other organs besides. Of all the different types of animal meat, offal is at the top of the list in terms of nutrients: it contains more vitamins and minerals per gram than any fruit, vegetable or steak-like cut of meat.

Of all offal, though, bone marrow and brain rank the highest in nutrient and calorie content. Now, brain is 60% fat, and bone marrow is almost pure fat. So, in terms of net return on calories, bone marrow is at the pinnacle, with brain close behind. These fatty foods would have comfortably satisfied the hunger of any hominin. But, besides the unparalleled caloric returns that they offer, bone marrow and brain are also among the most nutritious of foods for humans, with liver close on their heels. Sadly, the regular meals of fried liver and onions that our grandparents might have eaten are an unfashionable thing of the past. What we should think of as superfoods have instead become the forgotten foods of yesteryear.

Bone marrow and brain contain all the nutrients that were needed for the evolutionary development of the human brain – nutrients that plant foods could never have provided, no matter how much of them we consumed. This fact is not widely known, along with an even more surprising one: a diet of animal fat, offal and meat still provides all the nutrients that one needs to live healthily. In fact, we can live *more* healthily if we were to eat only these foods.

So, how did our *Australopithecus* ancestors come by these power-packed foods? By scavenging large-animal carcasses. However, scavenging under the nose of a big cat is risky and potentially fatal. Plundering leftover meat posed other problems too. First, raw meat becomes putrid after a few days, and a generous serving of bacteria with meat certainly wasn't the order of the day. Even modern-day chimpanzees, some of which are known to eat meat from time to time, avoid the decaying flesh of dead animals. Then there is the evolutionary state of our teeth to consider: *Australopithecus* didn't have the right dentition for chomping through raw fleshy meat. Plus they were a couple of million years away from using tools and so couldn't cut meat into smaller easier-to-chew pieces. Therefore, hominins must have decided that it didn't pay to risk life and limb merely for a few scraps of tough and rotten meat.

LONG SHELF LIFE

On the other hand, scavenging abandoned *bones* at kill sites certainly was worthwhile. Meat picks up harmful quantities of bacteria within twenty-four hours of exposure. In contrast, provided they are still contained within intact bone, bone marrow and brain stay fresh for several days, especially when they come from large animals. The long shelf life of marrow led early hominins to choose the safer strategy of focusing their scavenging efforts on bones. They could therefore afford to play a waiting game when accessing kill sites. A couple of days after stronger, fiercer and faster animals had had their fill and had finished rummaging around decomposing fleshy meat, the risk of being pounced on by predators would have been considerably lower. Moreover, bones were easily transportable. They could be carried off and safely dealt with well away from hazardous kill sites. The abandoned leftover bones made for easy pickings.

HITTING THE JACKPOT

But how to get the bone marrow and brain out of tough, unyielding bones? Even though early hominins couldn't make tools, they had a few other tricks up their sleeve. They figured out that they could use stones to smash bones (just as modern-day chimpanzees use stones to crack nuts) or let bones fall to the ground until they shattered. These simple strategies meant that the benefits of exploiting kill sites far outweighed the costs. Our ancestors had hit the calorie and vitamin jackpot.

Besides their nutritional and long shelf-life advantages, there were other benefits of bone marrow and brain. Existing on raw plant matter alone required constant foraging and a minimum of five hours chewing per day, whereas bone marrow and brain – being softer and easier to chew than any other part of an animal or fibrous plant food – were far more efficient choices. Plants not only have a low-calorie yield, but are also relatively difficult to digest. Brain and bone marrow, on the other hand, can be gulped down in a matter of minutes and are very easily digested.

How much time exactly did eating meat compared to eating vegetables save? A group of modern-day researchers decided to investigate and compared a diet of just root vegetables with a diet which was one-third raw meat and two-thirds root vegetables. They concluded that adding this much meat to the diet would have saved a whopping 2 million chews per year (roughly 5,500 fewer chomps every day). That's a big saving in chewing time and in chewing force. And, compared with fleshy meat, there would have been even greater reductions in chewing effort when consuming brain and bone marrow. What is the relevance of this? The time saved from not having to patiently chew for hours or continuously forage plant foods meant that there was now extra time to socialize and more time to develop tools. Less chewing also allowed *Homo* to evolve smaller faces and teeth – some of the features that were needed for the development of speech, another uniquely human trait. The bottom line is that the savings from chewing facilitated other brain-building activities and behaviours.

The soft spot that hominins had for bone marrow and brain endured over millions of years. It is only in the past few decades that these foods have gone out of fashion, in part owing to the outbreak of bovine spongiform encephalopathy (BSE) in British dairy cattle in the 1980s and its potential link to the human variant, Creutzfeldt-Jakob disease (vCJD). BSE was not just a tragedy for the animals and the people affected; it further damaged the reputation of offal. Naturally, since the BSE crisis a deep suspicion has emerged (both in the UK and elsewhere) about eating brain.

What is easy to forget is that the cause of the BSE outbreak was that manufacturers of cattle feed had added rendered remains of dead sheep to their cattle feed products (unbeknown to their farmer customers). It is thought that a prion protein that causes a similar brain disease in sheep called scrapie was able to leap from sheep to cattle via the adulterated feed. This serves as a cautionary tale about the dangers of imposing an unnatural diet on animals, one that evolution has not designed them for. And that warning should not be ignored when we consider making any changes to the diet that our own bodies, over millions of years, have evolved to eat. Fortunately, BSE has been virtually eliminated from British cattle. And not many know that despite the damage BSE caused to the reputation of British beef, the disease was almost entirely a problem of the dairy herd, not the beef herd. Perhaps this was because British beef is mainly grass-fed and beef farmers rarely use these manufactured products.

Because the best source of food for the evolving and growing human brain was a combination of bone marrow and brain, this diet paid off handsomely for hominins. Larger-brained individuals would survive to breed larger-brained children. *Australopithecus*, the predecessor of *Homo*, had a brain that was already 30% bigger than that of chimpanzees and other apes of comparable body size as far back as 3.5 million years ago. Besides having relatively large brains, the *Australopithecus* hominins were also bipedal and had human-like hand proportions. All the foundations that enabled hunting were being laid. Smashing bones put hominins on the road to an animal-food diet and humanity.

It is generally thought that the use of stone tools marked the beginning of human culture. The first-ever implements may have been very rudimentary, but they were revolutionary owing to the fact that stone flakes and other tools enabled our predecessors to butcher and cut meat. And, of course, 'stone' is one of the two concepts that the word 'Palaeolithic' points to (*lithos* being Greek for 'stone', and *palaios* for 'old').

As for processing food – we're not talking about pot noodles here, we're talking about cutting food up, which is also a form of processing (more on the benefits of food processing shortly). The sharp bone flakes produced during bone marrow acquisition may have inadvertently produced the first-ever cutting tools, and been in use a good few million years earlier than the development of stone tools. Hominins may also have used the bone flakes that are incidentally produced while butchering bones to manufacture shards for further butchery. If bone flakes rather than stone flakes were the first cutting edges used by hominins, there is a case to say that meat eating took off well before tool use and hunting. It may be that the 'The Bone Age' rather than 'The Stone Age' represented one of the biggest leaps forward in the evolution of mankind.

FROM TOOLS TO TEXTING

The importance of an animal-based diet in the evolution of the human brain is irrefutable. But there were other factors that also came into play to drive or support brain expansion: cognitive and behavioural stimulation, hunting, cooperation, food sharing, social interaction, gene mutations, increased infant dependency, the ability to sweat and outrun other animals, the reliance of the human baby brain on ketones, and the development of culture and language, to name but a few. The significance of these other influences is evident even today.

For example, when there is an absence of mother–infant bonding in childhood, the brain development of those infants suffers irreversibly; a lack of social interaction has serious consequences. But were social interaction, language and culture the main factors driving the growth of the human brain? If we look to other highly social primates, none of them has the cognitive skills of humans.

Social interaction on its own, therefore, cannot be the answer. Language and culture were also catalysts for brain expansion, but without animal foods providing the fuel for brain expansion, these factors alone could not explain the progress of the human brain.

And bizarre though it sounds, without our sophisticated ability to sweat we might not have progressed from scavengers to successful hunters. Most other species have the ability to cool down – dogs pant, vultures poop on themselves, elephants use their gigantic ears to dissipate heat. But it turns out that human sweat works better than the cooling systems of all other species; secreting water onto the surface of the skin is an extremely efficient way of dissipating excess heat. The upshot of our superior heat-loss strategy meant that we could chase our prey to death. Sooner or later our prey would have to stop to avoid overheating. We could keep going, at a slower pace perhaps, but for longer stretches, until our prey collapsed from heat exhaustion. We weren't fast – even goats are faster than us – but the 2 million sweat glands that we are born with gave us endurance.

Others speculate that it was our ability to grasp objects well that made the difference. Central to being able to grasp is the ability of the thumb to touch the other fingers (known as 'opposable thumbs'). In fact, many primates have opposable thumbs and can hold and manipulate even very small objects – having opposable thumbs was an adaptation to life in the trees when we needed to grasp branches. But when *Homo* began using stone tools more frequently, the human thumb became longer – relative to finger length – than any other primate thumb. Of all the primates, humans now have the best dexterity; we can grip heavy tools as well as text on the small screen of a phone.

So, yes, all these other factors did matter. Standing on two feet, being able to communicate, using tools, cooperating in the hunting down of animals, having opposable thumbs and the ability to sweat all enabled our ancestors to successfully add larger and larger amounts of animal food to their diet. You could even say that they were all game changers. But none of these other factors would have culminated in the massive human brain without the fuel of animal foods. And the more animal foods that hominins ate, the more their brains and intelligence grew. The more their intelligence developed, the more ingenious they became at acquiring even more animal foods. A positive feedback cycle was set in motion (see diagram opposite).

If we had continued eating as chimpanzees did then and still do now – that is, a plant-based diet – we would still have both the small brain and the limited intelligence of a chimpanzee, and a different gut anatomy. Some say that we are what we eat, but it's more accurate to say that we are the product of what our dim-and-distant ancestors ate.

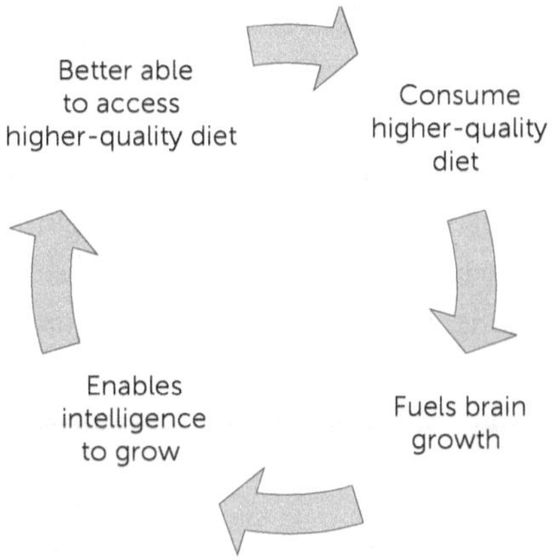

Diet initiated a positive feedback loop during human evolution.

Time	Hominin	Brain size / cm^3	Ape	Brain size / cm^3
3-4 million years ago	*Australopithecus*	440	Chimpanzee	400
2 million years ago	*Homo erectus*	930	Chimpanzee	400
300,000 years ago	*Homo sapiens*	1,350	Chimpanzee	400
Currently	*Homo sapiens*	1,260	Chimpanzee	400

Brain size in hominins and chimpanzees over the past 6 million years

During human evolution, of all the fauna available, large herbivores were the most beneficial to eat. Mammoths and other super-sized herbivores contained the greatest amounts of fat – something that our predecessors prized over all other foods. Animal foods also happen to contain more than ample amounts of the nutrients that are crucial for the development of the human brain, including docosahexaenoic acid (DHA), zinc, haem iron, choline, vitamin A and certain B vitamins. By contrast, some of these brain-building nutrients are difficult or impossible to obtain in sufficient amounts from a plant-based diet. The gradual change from a plant-predominant diet to one with increasing amounts of animal foods runs in parallel with the trajectory of our expanding brain over the past 6 million years. It is bizarre that the very same foods that evolutionary biologists consider to be of low quality (whole grains, lentils, seeds, fruits and vegetables) are exactly the foods that doctors and nutritionists misguidedly insist are the healthiest.

WHY PROCESSED FOODS WERE GOOD FOR US

Nowadays, 'processed foods' have a bad reputation. But the type of food processing that our ancient *Homo* ancestors did actually helped us evolve. What we are talking about here is pounding, cutting and cooking, which are all food-processing methods. Each one of these techniques makes the calories, vitamins and minerals in food easier to eat, digest and absorb, making the nutrition that's locked up in food more available, or 'bioavailable' as it is technically known. There is some debate, though, about one of these processing methods: cooking. This form of processing does, indeed, increase the bioavailability and digestibility of food. If *Homo* had dug up root vegetables or painstakingly split open grains and legumes before being able to use fire, he would have had to eat them raw. Besides the poor return in calories, certain uncooked plant foods are poisonous, and others actively detract from our health by reducing the absorption of other vital nutrients. (We'll further explore in detail the reasons why a plant-based diet is a bad idea – whether we cook those plants or not – throughout the book.)

However, cooking depends on the ability to control fire. While *Homo* may have 'cooked' in a very rudimentary fashion as long as 700,000 years ago, most anthropologists agree that it is much more likely that fire was only routinely used some 300,000 years ago, by which time the *Homo* brain had already grown to its maximum size. Cooking, therefore, doesn't account for the tripling in size of the brain over the previous few million years. Besides, raw meat is digestible, has a good return on calories, still contains bioavailable nutrients and when eaten fresh is unlikely to contain toxic substances. Furthermore, brain and bone

Comparing raw plant food-eating chimpanzees with cooking carnivores.

As a raw plant food consumer, the average chimpanzee needs to spend an average of six hours a day eating six pounds of plant food. Tedious.

As a cooking carnivore, the average human needs to spend an average of fifteen minutes a day cooking and an average of fifteen minutes a day eating a pound of fat, offal and meat. Tasty.

marrow, the very first animal foods that our ancestors started eating, were soft, extremely easy to digest and did not require cooking. This, in turn, allowed *Homo* to evolve smaller chewing muscles and less sharp teeth, which made more space for speech and brain development.

WHY AREN'T OTHER CARNIVOROUS ANIMALS AS CLEVER AS WE ARE?

A question that often crops up in any discussion about the role of meat in the evolution of intelligence is why other carnivorous animals didn't evolve in the same way. If adding meat to our ancestors' diet was responsible for their brain growth, how come all carnivores aren't as clever as man? A good question, and one we can help answer with a little thought experiment.

Imagine swapping brains for a day with your pet dog or cat (real or imaginary – it doesn't matter, as this is a thought experiment). While you are preoccupied chasing squirrels or mice, your pet's newly installed brain now understands many things. Fire, for instance: how to make one with kindling, sticks and logs; how to light it with sparks struck from flints; its power to cook meat to make it easier to eat. Your pet would also understand tools – the usefulness of a stone or a flint broken in such a way that it yields a sharp cutting edge that can be used to cut meat from a carcass, or to cut wood to make a fire. This clever pet would understand that pointy-ended quartz flakes can be attached to lengths of wood to make spears – weapons that can be used to hunt large prey. But while your pet now understands these things, and more, it gains no immediate advantage, and no *evolutionary* advantage, from that knowledge. Why? Because dogs, cats or any of the great predators that lived alongside us during our evolution lack arms, hands and opposable fingers and thumbs. They are unable to turn intelligence into anything practical.

Of all the meat eaters, the growth of the hominin brain occurred in an animal that was already bipedal, with upper limbs that were free to manipulate the world around it, controlled by its ever-growing intelligence. With its hands, our ancestors could turn theory into practice. The evidence in the anthropological records shows that *Homo erectus* was making and using stone chopping tools 2.6 million years ago and more sophisticated tools such as hand axes and other large cutting tools 1.7 million years ago. With a precision grip afforded by opposable thumbs we had the facility to turn our meat-fuelled intelligence into tools; tools that gave us huge evolutionary advantage.

You can now swap brains back. And put the squirrel down.

The other argument used to question the importance of meat in our diet is that we don't have the teeth of carnivores and therefore aren't evolutionarily adapted to hunting and eating meat. But we never needed to evolve large fangs, or, indeed, claws, because that's not how we hunted. We let other predators do the hard work, bringing down and killing prey – then we nipped in and scavenged the carcasses. Later we developed weapons that enabled us to cut out the 'middleman' and hunt for ourselves. Spear heads mimicked large teeth and we used sharp tools to slice meat and blunt tools to pound meat. As a result we didn't need to rely as much on our teeth as other carnivorous animals do. And incidentally, gorillas and our other herbivorous primate cousins have much bigger canine teeth than we do, but their function is primarily social display, not carnivory.

Another evolutionary adaptation that is unique to us among our ape cousins is our shoulder anatomy, which gives us a superior ability to throw things. While launching a spear may not be critical to survival today, being able to throw well was critical in human evolution – it provided protection against predators and allowed us to hunt more successfully than any other species. All other apes have shoulders that are well adapted to swinging from trees; the human shoulder, on the other hand, has evolved to throw with accuracy and speed.

Uniquely to intelligent hominins, arming ourselves with chopping stones, then hand axes, then spears and arrows, meant we progressed to more and more sophisticated ways to access a high-quality meat-, fat- and offal-rich diet. But one outcome of developing intelligence we did share with other carnivores was social skills. We learned to cooperate in hunting groups, which were more effective in taking down large animals than individuals. We share this sort of social intelligence with other predators, such as dogs that hunt in packs. Another human evolutionary adaptation was the development of white eye sclera, which aided communication and cooperation during group hunting (by making it easier for others to detect the direction of gaze). Wolves hunt in a very similar way to humans, and though they don't have white sclera, they do have eye markings that aid visual communication between pack members.

BRAINY FATS

The composition of brain tissue is remarkably similar in all mammals. However, species that have higher levels of complexity in their brains, or those with relatively larger brains, have a greater dietary need for two key fats: docosahexaenoic acid (DHA) and arachidonic acid (AA). Wild plants that were part of the African savannah during our evolutionary history contained virtually no DHA or AA, and fleshy meat contains only moderate amounts. On the other hand, offal, especially the brains of grazing animals such as buffalo and bison, is an excellent source of both these critical fats. But, because mammals have a limited capacity to make these key fats, consumption of pre-formed DHA and AA was essential to promote brain growth during evolution. Bone marrow and brain from the large herbivores roaming on the African savannah provided ample vitamins, minerals and calories, as well as the crucial fats DHA and AA necessary for our evolving brains.

WAS FISH CONSUMPTION RESPONSIBLE FOR THE EVOLUTION OF THE HUMAN BRAIN?

Some scientists have suggested that the consumption of fish was critical in the evolution of the human brain. This idea is backed up by the fact that 30% of the dietary protein of humans in Western Europe came from marine mammals and that fish is an excellent source of brain-essential nutrients. But there is a problem with this thinking. The regular consumption of marine foods in Africa and the migration of modern humans to Europe are both relatively recent events (roughly 40,000 to 90,000 years ago respectively) and by which time the phenomenal brain growth in *Homo sapiens* had already taken place. Besides, as already mentioned, land-based animals provided more than ample amounts of DHA and the other brain-essential nutrients. In fact, the brains of grazing animals were by far and away the best source of DHA. Plus, as we will shortly see, the calorie return on fishing compared to hunting large fatty land animals made fishing much less worthwhile. There is, therefore, little basis for the suggestion that fish consumption was vital for the evolution of the human brain or that the terrestrial dietary supply of DHA or other nutrients was a limiting factor in the expansion of the hominin brain.

Food / 100g	DHA / mg	AA / mg
Wild tubers/roots	0	0
Mixed wild plant foods	0	0
African ruminant brain	861	533
African ruminant liver	41	192
African ruminant muscle	10	152
African ruminant fat	Trace	20-180
African fish	549	270

DHA and AA (brain-essential nutrients) in plant and animal foods.

WHY A PURE LEAN-MEAT OR A PURE PLANT-FOOD DIET AREN'T GOOD IDEAS: THE PROTEIN CEILING AND THE PLANT-FOOD CEILING

All our calories come from either protein, fat or carbohydrates (known as 'macronutrients'). There is a biological limit to how much protein our hominin ancestors could (and modern-day humans can) consume. The upper limit on protein consumption, known as the 'protein ceiling', is of the order of 50% of calorie intake, which implies that the remaining 50% of calories must be derived from fat and/or carbohydrates. Not many know that eating a diet of pure protein for several weeks is dangerous and can even cause death. This is because excessive amounts of protein overwhelm one's liver and kidneys, leading to dangerously high levels of ammonia and urea in the blood.

In fact, early American explorers who consumed only wild rabbit meat ended up dying of protein poisoning – this has come to be known as 'rabbit starvation'. Some of you might be thinking that surely rabbit meat can't be harmful, as it's a type of meat? The problem for the American explorers was that rabbit meat is very lean, so they were eating almost pure protein. Another reason why eating too much protein causes problems (if this is carried out for several months or longer) is because the process of converting protein into glucose requires a large amount of energy. And when more than 50% of calories are supplied by protein, this leads to a negative energy balance, i.e. more energy is used than is produced. If the intrepid explorers had managed to get hold of fatty meat they would most likely have survived. The bottom line is that a diet in which protein is consistently the main source of calories is not a viable long-term option.

One of the key differences between *Homo sapiens* and certain carnivorous animals (such as the cat family) is that the latter need to eat much more protein; humans obtain roughly 15–20% of their calories from protein, whereas for cats the figure is 50%. This is because cats must convert large amounts of protein into glucose to fuel their brains. The human brain, though, does not need large amounts of glucose because it is able to use an alternative fuel – ketones – which is readily supplied by a high-fat diet. This ability of the human brain to use ketones was critical in our evolution (more on ketones in chapter 11).

Because of the biological limit on protein consumption, eating just lean meat (which has a very high protein content) would have been a problem for our hominin ancestors; they would have had to make up their calories with either fat from animal foods or carbohydrates from plant foods. Surprisingly, though, there is also a restriction on plant-food intake, which is determined by the colon's capacity to process fibre. Most modern plant foods have been bred to be much less fibrous than they used to be before the advent of agriculture (10,000 years ago), making the fibre limitation an even bigger concern for our hominin ancestors than it is for modern man. Furthermore, the presence of toxins and anti-nutrients in plant foods (such as phytic acid, trypsin, tannins, alkaloids and protease inhibitors) reduces the absorption of certain vitamins and minerals. Plant foods thus present a double whammy: they have lower levels of nutrients than animal foods do and they also *impede* the absorption of nutrients from animal foods. These are some of the features of plant foods that deem them as having poor 'bioavailability' (poorly absorbed by the body). In contrast to plant foods, the nutrients in animal foods are easily absorbed and have excellent bioavailability.

The human brain is a power-hungry organ: it uses ten times as much energy as other organs do. Our brains account for only 2% of our body mass but use 20% of all the calories we consume. Compared to most other animals, the human brain uses prodigious amounts of energy; for example, our brains use twice as many calories as chimpanzee brains and three times more than squirrel and mice brains. In the entire animal kingdom, the metabolic 'cost' of the human brain is one of the highest – it's an extremely energy-expensive organ. Where did our expanding brains get their extra calories from? It was animal fat that delivered the calorie needs of our growing brains.

The advantages of fatty animal foods over plant foods can be summarized as follows: higher levels of calories, vitamins and minerals; correct amounts and types of protein and fat; better bioavailability of vitamins and minerals; no toxic substances; no anti-nutrients; easier to chew; no fibre and easier to digest.

BETTER RETURNS FROM LARGE PREY

Our ancestors habitually scavenged and later hunted for prey larger than themselves. Let's now take a look at why they went for large prey by comparing the calories that one can get from different animals and foods. To start with, just the meat of a woolly mammoth has 3.6 million calories, an auroch has about a million calories and a bison has 600,000. Relatively speaking, large animals contain about 40% more fat than smaller animals, and the bone marrow yield from large animals is even higher (if we look at modern-day animals, a cow is 35% fat, while a rabbit is only 3% fat). The calories available in mammoths, aurochs and bison would more than double if we were to include their fat. That would be an astounding *7 million* calories per mammoth. To get the equivalent calories from other foods would entail hunting roughly 3,000 rabbits or catching 6,000 fish or finding 70,000 eggs or gathering ten tonnes of fruit and vegetables. Clearly, it would take far less time and effort to kill one woolly mammoth than it would to kill several tonnes of rabbits or fish, or to find thousands of eggs, or to pick a few tonnes of fruit and vegetables.

Food type	Net calorie yield per hour / kcal
Large game	15,000
Small mammals	3,800
Seeds and nuts	800-1500
Roots and tubers	400-1400

Calorie yield from hunting and foraging.

Some scientists have gone so far as to work out the calorie return per hour of obtaining these different foods – there's a tenfold difference between large game and plant foods. No wonder our ancestors went for large game first rather than fish or bunnies. Large grazing animals were critical to human survival during the Pleistocene era (between 2.6 million and 11,700 years ago). Unfortunately, this strategy may have inadvertently caused the extinction of many large animals. The fact that early humans were highly skilled hunters, and could hunt animals such as elephants and mammoths at will, thus paid 'fat' dividends for mankind.

A DIGRESSION INTO THE NEANDERTHAL MYSTERY

Much of this chapter has been about using evolution to explain diet; now, let us use diet knowledge to explain evolution: specifically, the advantages that *Homo sapiens* had that might have resulted in its survival, while Neanderthals became extinct.

At first glance, it seems counter-intuitive that the long-established indigenous Neanderthals should die out while the new arrivals prevailed: the Neanderthals had survived the extreme climatic instability (fluctuations between cold glacial and warmer interglacial phases) that was the norm in their habitat, Eurasia, for much of that time; and there were not that many humans on the planet overall (as few as 10,000 at certain times during the Pleistocene era). *Homo sapiens*, who evolved in Africa, was a relative newcomer to the harsh Eurasian habitat, arriving some 50,000 years ago rather than the Neanderthals' 500,000; they would not have had the highly effective subsistence strategies that the Neanderthals clearly had. The reason for the success of *Homo sapiens*, though, could be down to diet.

There must have been intense competition for the dwindling population of large game in Eurasia after *Homo sapiens* took up residency. The *Homo sapiens* interlopers were tall and lean compared to the short but large-bodied Neanderthals, who would have been less adept at hunting smaller, faster-moving game; therefore, the Neanderthals would have been forced to eat more plant foods as the big game disappeared. Compared to game, plant foods have a much lower calorie content – there is therefore a limit on how well plant foods could have compensated for a reduction in the consumption of large game, especially considering that Neanderthals had greater energy needs than *Homo sapiens* due to their heavier bodies. The fossil record from the teeth of Neanderthals supports this theory: it clearly shows that they suffered high levels of nutritional stress after the arrival of *Homo sapiens* in their territories.

Dietary iodine could provide more potential explanations. Firstly, humans have a genetic alteration that improves our ability to extract iodine from food; Neanderthals did not possess this gene. Our superior iodine-absorbing capacity meant that meat and offal could provide all the iodine that we needed. Moreover, although both species' diets were remarkably similar, one difference was that the Eurasian *Homo sapiens* was known to have consumed aquatic (marine and freshwater) resources as well as land animals. Some scholars have suggested that it is the high iodine content of marine fish that may have given *Homo sapiens* an advantage over Neanderthals. The reduction in large game, therefore, would have affected iodine levels in Neanderthals much more than it would have done in *Homo sapiens*.

The idea that the Neanderthals were iodine-deficient is strongly corroborated by the conspicuous similarities between their skeletons and the skeletons of humans with iodine deficiency, a deficiency which causes distinctive physical deformities, brain damage and learning difficulties. In fact, Neanderthal skeletons resemble those of iodine-deficient skeletons more closely than they resemble those of healthy humans.

It could be that an accumulation of these small dietary advantages gave us the decisive edge. That *Homo sapiens* were superior at absorbing iodine, consumed marine fish and were adept hunters of smaller game all make for a plausible case explaining why *Homo sapiens* prevailed over Neanderthals in Eurasia.

BACK TO THE PRESENT – DOES BRAIN SIZE EVEN MATTER?

Brain size during evolution correlated with cognitive ability: as our brains increased in size, so did our intelligence. However, in most animals, brain size correlates with body size: that is, the larger the animal, the more likely it is to have a large brain – and vice versa. Thus, in absolute terms, elephants have the largest brains of all land animals and sperm whales have the largest brains of all animals. So, you might ask, why aren't elephants and sperm whales at the top of the food chain? It turns out that relative rather than absolute brain size is what matters. But, besides the relative size, there are likely to be many more known and unknown factors that matter. For example, the complexity (how convoluted a brain is) and how asymmetrical a brain is also matter (strangely, having a lopsided brain is a sign of cleverness).

WREAKING HAVOC ON THE PLANET

The fact that early humans were highly skilled hunters, and could hunt animals such as elephants and mammoths at will, was critical for the development of humanity. But as a result, we hunted down all the megafauna, such as mammoths, one by one to extinction. We also brought about the extinction of large carnivores by out-competing them for prey with our increasingly effective hunting methods. Our superior intelligence has been wreaking havoc on biodiversity and the ecosystem for a very, very long time.

The good news is that we now know how to farm animals in a way that not only benefits the ecosystem but also helps with carbon emissions – how this is possible is the subject of the final section of this book.

BASIC INSTINCTS

Hunger is a basic instinct common to all animals. At a fundamental level, the objectives of eating are to satisfy hunger, obtain energy and enable procreation. In times of fluctuating food availability, Palaeolithic man would not have stressed about the types of food they were eating: if their first-choice foods weren't available, they would have made do with anything that was edible. Unlike modern man, our ancestors were far less likely to have been burdened with the guilt of consuming unhealthy foods or eating other living creatures.

In sharp contrast to those prehistoric times, modern humans often battle with psychological and moral conflict about what they should eat. We worry about how what we choose to eat affects our health, our appearance and our conscience. However, the fundamental biological need for the diet of our ancestors has not changed. The human body and brain have evolved owing to a certain way of eating, and so the diet that suits humans has been built up over millions of years. The very same foods that made humanity possible are the foods that yet suit us best. Ethical issues aside, it might be naïve and possibly even arrogant to think that we are now able to come up with a diet that is superior to the one through which we evolved.

In conclusion, we didn't evolve to eat meat, we evolved *because we ate meat*. Our different diets, our different digestive systems and the different ways that we obtained our food were some of the most critical factors that caused us to part from our chimpanzee cousins. Our brains could only have developed because of what our predecessors ate during our evolution.

But somehow, sometime, somewhere, it all went horribly wrong. The mistakes started when we began to flirt big-time with grains and cereals after taking up agriculture ten thousand or so years ago. Some scientists have even suggested that this transition in our diet may have contributed to the recent reduction in our brain size. Neolithic farmers can be forgiven for being unaware of our evolutionary past, and of the implications of experimenting with new ways of eating and of growing food, even if doing so suited the changing social orders emerging at the time. But we shouldn't forgive recent dietary research. Seemingly ignorant, or dismissive, of our evolution, it has managed to misinform dietary guidelines – guidelines that are just about the complete opposite of our ancestral diet.

We'll see how this came about in the next chapter.

CHAPTER 3

SCIENCE FICTION

A guide through the minefield of nutrition science

'Facts are stubborn, but statistics are more pliable.'
Mark Twain

A NEW DRUG IS ANNOUNCED BY A PHARMACEUTICAL company. They make bold claims for it: *this will protect you from life-threatening diseases and you will feel healthier.* It has to be taken in large doses, three times a day, every day for life. It works by affecting every cell in your body, every metabolic process that keeps you alive and every organ, including your brain. Yet while the promises made for the drug sound wonderful (and they've also grabbed the attention of the media), you feel anxious about whether there are risks associated with something that has such a profound impact on all parts of your body. So you ask: has it been rigorously tested?

Oh yes, says the pharmaceutical company. We did a trial. First, we found 1,000 volunteers who had the time and inclination to take part. They were each sent a large supply of the drug. We asked them to help themselves to a serving of the drug three times a day (it has been designed to be taken orally). We trusted them to self-dose correctly – neither too little nor too much – and to take those doses three times a day. We trusted them not to mix the drug with alternative drugs widely available on the market (and which their bathroom cabinets were likely full of). We are confident no one cheated or deviated, because we asked them, and they all said, 'No we haven't, honestly, we haven't.' After a month we asked them to report how they felt, and we took some blood tests, and compared the results with those from the beginning of the study. We found a 20% increase in the number of people who said they felt healthier, and this agreed with the results of the blood tests. What do we mean by 20% increase? You'll have to read the small print. Were there any side effects? Funny you should ask, but we observed that quite a few people gained weight, and others developed a hormonal imbalance, but we don't believe these were owing to the drug. Do we need a licence? That's the great part – no!

We hope that most people would not adopt a drug – one that required taking significant quantities every day for the rest of their lives – where the testing of the drug was carried out in poorly designed trials which relied on self-administered doses and self-reporting of results.

Yet, replace 'drug' with 'diet' in the above paragraphs, 'bathroom cabinets' with 'kitchen larders and fridges', 'pharmaceutical company' with 'dietary researchers', 'gained weight' with 'became obese' and 'hormonal imbalance' with 'type 2 diabetes', and that's just what people do – they change what they take into their bodies three times a day, every day, year upon year, and all because of dietary studies they've seen reported in the media.

From the earliest cave art depicting hunters killing animals, right up to present-day headlines shouting out at us to 'eat five-a-day to live longer', humans have long realized the importance of food. Even so, many are unsure as to what exactly constitutes a healthy diet. Unlike wild animals, who instinctively know what to eat, we humans appear to be confused and often end up falling for newspaper headlines telling us what we should or shouldn't consume. This confusion is compounded by the media changing their tune at the drop of a hat. For example, in the *Daily Mail* in 2012, it was claimed that 'two cups of coffee a day could help relieve shakes caused by Parkinson's disease'. In 2017, the opposite appeared to be true: 'Coffee does NOT treat Parkinson's'. Until, that is, the following year, when things changed again: 'Coffee could combat Parkinson's'.

As for the advice on eggs, this has flip-flopped at least once every decade during our lifetime. In the 1960s, eggs were viewed as healthy (epitomized in the Dairy Board slogan 'Go to work on an egg'); in the 1970s we were told that eggs are full of cholesterol and should be avoided; in the 1980s we were told that cholesterol in our diet doesn't have a big effect on blood cholesterol and it was fine to eat a few eggs. Every new finding from the world of nutrition science is sensationalized by the media with impunity.

The list of U-turns – not only in the media, but also in the official dietary advice – is long. Trans-fat laden margarines have gone from being 'heart healthy' in the 1970s to the devil incarnate in the 1990s. Salt is now the grim reaper's scythe. Eggs have experienced boom, bust and revival. We sometimes wonder whether we should believe anything at all that we read about food, even though all these assertions are backed up by very precise and apparently credible statistics. If you were wondering whether to eat eggs or not, you can try typing 'eggs, diet and health' into Google Scholar (where publications from academics can be found) to help you decide. What you'll get is over a million results. How on earth are we supposed to know what to eat? Welcome to the confusion that is nutrition science.

But it isn't only nutrition that is beset with contradictions. There are countless examples of major about-turns in other public policies too. Some strategies have turned out to have been a complete waste of time and money, or to have damaged the environment. Part of the problem lies in the fact that many of us trust experts and blindly believe advice from politicians. So it was that a decade ago we were told that diesel is better than petrol for the environment. Encouraged by cuts in diesel fuel duty by governments, we bought diesel cars. Then we learned that politicians had been misled, in part by car manufacturers, about the benefits of diesel. In fact, diesel was worse than petrol in terms of particulate and nitrogen oxide and nitrogen dioxide emissions. And so, government performed a U-turn. Diesel was now out and petrol was back in.

Another example of a government U-turn is the planting of Norwegian spruce trees across large swathes of the UK after the Second World War. It was later realized that Norwegian spruces are challenged by droughts, and also by pests present in the UK. In due course we sensibly started replacing the non-native spruce with native trees instead. Also after the war, the government granted farmers financial incentives to remove hedgerows. This disastrous policy caused immeasurable damage to ecosystems. Now governments give grants to plant and protect hedgerows.

These examples parallel, in many ways, the current situation with dietary advice. In the 1970s we were told that cutting out butter, red meat and eggs, and replacing them with wholemeal grain foods and fruit and vegetables, would improve our health. Our parents' generation thought such a diet was a recipe to put on weight, and they were right. The dramatic rises in diabetes and obesity are testament to the deterioration in our health over the last few decades. Politicians in Westminster, and institutions like the NHS, have believed poor-quality information as told to them by nutrition scientists. As a result we have been misadvised to fuel our bodies with glucose rather than animal fats.

This chapter is a story of irresponsible science within the nutrition world. Many today believe that the dietary recommendations that came into existence fifty years ago are responsible for the subsequent worldwide rise in obesity and diabetes. Furthermore, as diabetes is a risk factor for dementia, cancer and many other diseases, these dietary recommendations may be behind many Western diseases. A grave and bold accusation, indeed. To be fair, nutrition scientists didn't set out to deliberately deceive us; they must have believed their advice was correct. But it's to be hoped that we will come to see the current nutrition guidelines for what they really are – a set of ill-thought-out recommendations that should never have seen the light of day. Here, we set out why the current nutrition guidelines are seriously flawed and in need of drastic change.

EPIDEMIOLOGY

At the heart of the problem is that the current dietary advice comes from the field of epidemiology. Epidemiological studies give us information about who has a disease, where they got the disease and when they got the disease. And if we look at what those diseased people do – such as what they eat on a daily basis or whether they exercise or smoke – we often find certain risk factors that are associated with their diseases.

Epidemiology can, indeed, be a very useful science. For example, in the landmark 'smoking and lung cancer' study from 1950, a group of 709 men with lung cancer were compared to an equal size group of men without lung cancer. The study found that those who smoked twenty-five cigarettes a day were fifty times more likely to get lung cancer than the non-smokers. You don't need to apply complicated statistical methods to see the overwhelming association between smoking and lung cancer in this study. Despite this remarkable association, the authors of the smoking and lung cancer study added this caveat: 'This is not necessarily to say that smoking causes carcinoma of the lung.' This is because even though this study showed that there is a very high chance that smoking *might* cause lung cancer, *no* epidemiological study can definitively prove a cause.

Interestingly, the authors of the smoking and lung cancer study (Sir Richard Doll and Sir Austin Bradford Hill) also found that smoking was associated with a twofold increased risk of heart disease in smokers. Unlike most of the authors of modern nutrition studies, Doll and Hill had high standards and went on to say that a twofold increased risk wasn't a strong enough association to imply that smoking *causes* heart disease.

Here's an extreme scenario to show why an association found in an epidemiological study does not necessarily mean causation. Imagine that we looked at the habits of two groups, each containing fifty people. Both groups were a bit strange. The first group, who were selected from the Russian Roulette

Association ('new members always welcome'), didn't consume any fruit or vegetables. The second group, volunteers from the Health and Safety for Fun and Profit Association (motto: 'Don't do anything rash'), ate prodigious amounts of fruit and vegetables. After five years everyone in the first group had died, whereas only three people in the second group had passed away. Conclusion: in this study, eating fruit and vegetables was associated with a decreased risk of dying.

We hardly need to explain what's wrong here.

Now an example of a real-life study: 'Egg consumption and risk of type 2 diabetes in men and women', published in 2009. The participants of the study had previously supplied baseline data as to how often they ate eggs (during the year before the baseline). With that information available, the study checked the men (on average twenty years post-baseline) and the women (on average eleven years post-baseline) to see whether or not they had developed diabetes. (Among many things that were unclear, the paper didn't explain the different follow-up periods for men and women.) From this it was concluded that those people who ate seven or more eggs a week in the year before the start of the study had a 58% higher risk of developing diabetes after ten years.

There are many questions or challenges we might make regarding this study and its conclusion, such as:

- Can we really rely on people accurately recalling how many eggs they were eating per day in the past twelve months?

- Did the researchers consider what else people were consuming – foods that might reasonably be expected to be taken alongside eggs? Whole-fat milk for instance? This question was included in the baseline survey. Interestingly, milk consumption increased as egg consumption increased (as did smoking), more than trebling from the low or no egg consumption group to the seven or more a day group. Was it milk (or smoking) that was associated with the incidence of diabetes? See the figure below, produced using data from the paper (the researchers offered no figures of their own). The paper didn't think these other associations worth noting or discussing.

- What happened to the diet of the volunteers during the eleven to twenty years after the baseline measurement? Did they continue eating the same number of eggs for the next one or two decades? What are the error bars in the baselines? Again, no comment by the authors.

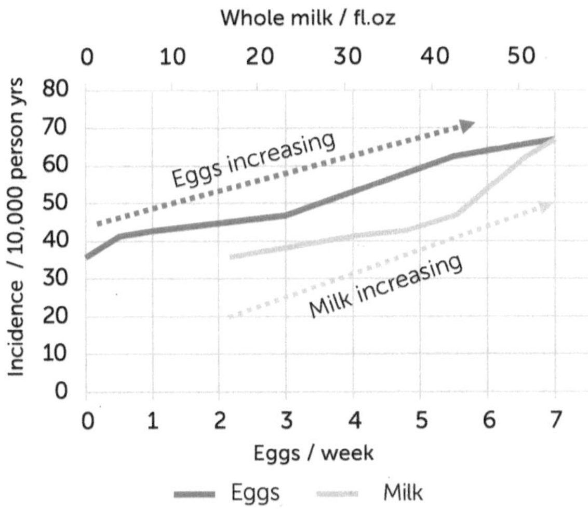

Type 2 diabetes incidence in men versus eggs and whole milk consumption.

And when the conclusion says '58% increase in risk', that's a *relative* change in risk – but what was the change in *absolute* risk? Relative and absolute risk portray very different pictures, as we shall shortly see. We're not insinuating that epidemiological studies deliberately set out to mislead. In fact, researchers in the field would almost certainly endeavour to account for all possible factors that might be playing into the results they measure, in order to strengthen their conclusions. But epidemiological studies are never closely controlled experiments, where one factor, and only one factor, might be varied, such that the resulting outcome (if there is one) can be directly linked to the factor that was studied and to nothing else.

ABSOLUTE AND RELATIVE RISKS TELL VERY DIFFERENT STORIES

Let's look at absolute risk and relative risk a bit more closely. The following figures illustrate how a small increase in *absolute* risk can be magnified enormously when the same figures are stated as a *relative* risk.

In the figures overleaf, the increase in relative risk from partaking of a witch's brew is indeed 20%, as reported. The research observed an increase in the chance of skin eruptions rose from 5 people in every 100 to 6 people in every 100. And 6 is 20% bigger than 5. So, yes, there was a 20% increase in *relative* risk.

Focusing now on *absolute* risk, in the second figure below, the change from 5 people in every 100 to 6 people in every 100 is one extra pimply person. Just a

'New research – adding eye of newt to your broth increases likelihood of boils by 20%'

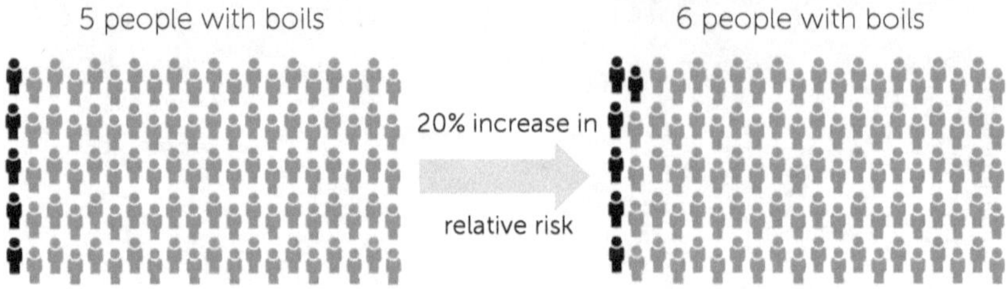

Newspaper headline, early edition.

'Researchers slammed for wasting funding – adding eye of newt to your broth increases likelihood of boils by only 1%'

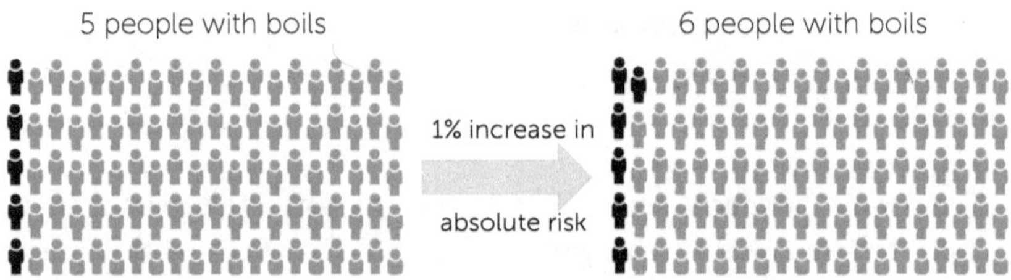

Newspaper headline, later edition.

1% increase. Same data, different story. No wonder journalists are tempted to go with the 20% increase in relative risk rather than the 1% increase in absolute risk.

The media take their cue from the original scientific papers – scientists know full well that their results sound more impressive when they report the relative risk. The difference is that the researchers clearly understand the difference between absolute and relative risk, whereas the average *Daily Mail* journalist most likely doesn't.

The figures used in this example are similar to a real study that was carried out to look at whether the chance of developing bowel cancer was related to increased processed meat consumption. The risk of developing bowel cancer is thought to be 5.6%. Consuming 50g of processed meat per day (for example, sausages or bacon with additives, such as nitrites) was found to increase the risk to 6.6%, giving an absolute risk increase of 1%. But what hit the headlines was the 18% change in relative risk, because it sounded more shocking. Even the risk-averse members of Health and Safety for Fun and Profit might consider that a 1% increase in risk is acceptable. Especially if they were partial to a full English breakfast. Others, though, might be thinking that even a 1% increase in risk is not acceptable – but we will now explain why even the 1% increase in risk is an unreliable figure.

MISSING: TWO BIG MACS

Misreporting of food intake in nutrition studies is a big problem in epidemiology. Failing to obtain accurate accounts of what or how much study participants eat clearly undermines how much nutrition studies can tell us with certainty. This comes about because most people, consciously or unconsciously, fail to be completely honest when it comes to revealing what they consume.

The scale of the problem has been demonstrated by a study carried out by a team at the University of Essex. They asked a group of 221 adults to keep a food diary over a period of time. The researchers used a technique ('doubly labelled water') to measure the calories that were being consumed while the food diaries were being kept. This 'doubly labelled water' method is 100% accurate at measuring calorie intake. What the researchers found was that, on average, the participants under-reported how much they ate by around 900 calories per day. We're not talking about an apple or two, but the equivalent of nearly two McDonald's Big Macs. And the study found that, contrary to what is generally believed, it wasn't only the overweight people in the group who under-reported their intake; everyone misplaced hundreds of calories. It appears to be an unavoidable human characteristic that we struggle to be honest, or are forgetful, about how much we eat, when asked. The same 'economy with the truth' is evident if we're asked how many units of alcohol we drink every week, or about the number of sexual partners we've had. This self-deceiving facet of our behaviour is well known.

The bottom line is that an epidemiological study is not an experiment. In epidemiological studies, researchers make observations about the result of things that their participants claim they have done. In experimental studies, researchers introduce a specific intervention and then measure the effect of the intervention. What then is the point of epidemiological studies if they can't prove anything definitively? Well, some epidemiological studies can generate useful hypotheses, as in the landmark smoking and lung cancer study by Doll and Hill in 1950. They observed that smoking twenty-five cigarettes a day was associated with a fiftyfold increased risk of lung cancer. This study was well designed and the association that it found was so strong that it clearly suggested there was a link between smoking and cancer that deserved further attention and study. The trouble with most nutrition studies is that they usually only result in weak associations. And when these weak hypotheses are tested in controlled trials, they often turn out to be just a hypothesis, and not a cause.

Another avenue to confirm a hypothesis from an epidemiological study is to look at whether there is a biological mechanism to explain the hypothesis. After the smoking and lung cancer study, various chemicals in tobacco (such as 'polycyclic aromatic hydrocarbons' and 'tobacco-specific nitrosamines') were in the spotlight. And these have been shown conclusively to promote tumour growth. In this case the strong association found in the study turned out to be an actual cause.

A further problem with nutrition studies is that it is almost impossible to make only one change in a study. Imagine a study where a group of people were told to cut down on their meat consumption and they were compared to another group who continued eating the same amount of meat. If the group who ate less meat didn't change anything else in their diet, they would end up eating fewer calories (the calories that were in the meat that they have cut out). If they substitute the meat with another food (for example, more vegetables) to make up for the calories from the meat that they have cut out, then they will be eating less meat but more vegetables, i.e. there will always be two differences in this study. If it were found that the group who ate less meat lived longer, one couldn't say with certainty whether it was because they ate less meat or whether it was because they ate more vegetables. This is why it is nearly always difficult to make only one change in a nutrition study.

BUT NOT ALL NUTRITIONAL STUDIES ARE USELESS

Coeliac disease is an autoimmune disease which mainly affects the bowel and is triggered by foods containing gluten. Symptoms of coeliac disease include diarrhoea, stomach pains, fatigue, weight loss and skin rashes. In children, faltering growth is an important feature of coeliac disease. But in the 1930s, the cause of coeliac disease was not known. However, a Dutch paediatrician practising at the time, Dr Willem Karel Dicke, noticed that his patients' symptoms improved with certain diets. The astute Dr Dicke had good intuition, superb clinical knowledge and an excellent scientific approach. He hypothesized that wheat-containing foods made his patients' symptoms worse.

Between 1934 and 1936 Dr Dicke encouraged his patients to experiment with different diets, including wheat-free diets. And when he heard from a young mother that her coeliac child's rash improved rapidly when bread was removed from the diet, Dr Dicke took her testimony seriously. Nowadays, sadly, many in the medical establishment would be dismissive of such a comment from a patient. In Dr Dicke's mind, this confirmed that eating wheat was a trigger for coeliac patients. Dr Dicke conducted more trials and later wrote a thesis on wheat and coeliac disease. Following his careful observations and experiments we now know that wheat is, indeed, the offending food in coeliac disease, and that it is the gluten in wheat that's the culprit. Looking at the underlying biochemical mechanism confirms this theory. Whereas in the earlier example, there is no biochemical mechanism to confirm or refute whether eggs cause diabetes.

How is all this relevant to modern-day studies? Dr Dicke arrived at his new theory by first noticing something very unusual and then investigating the idea in a sensible and scientific way. If all nutrition trials took the approach that Dr Dicke did, we wouldn't have the bizarre and contradictory dietary headlines in the media every other day. Most modern nutrition studies are beset with inconsistencies; they are usually epidemiological studies which rely on people remembering their food intake and come up with only weak correlations. In these kinds of studies, it is very difficult to prove that a particular food causes a particular disease. As one professor of medicine and data science at Stanford put it, nutritional science is 'a scandal' that should 'go to the waste bin'.

RANDOMIZED CONTROL TRIALS – THE GOLD STANDARD OF TRIALS

In a randomized control trial, a sample of people drawn randomly from the general population are assigned to either a control group, or to a 'treatment' group, the treatment group being the people that are going to receive the intervention in question (for example, a specific change in their diet). At the outset, both groups would be assessed to establish a baseline, the starting point against which results are compared. At the end of the study period, and possibly at other intervals after the start of the trial, the groups would be retested to see how they had responded to the intervention, in particular how the treatment group compared to the control group.

To eliminate confounding factors, each group should have the same range of alcohol drinkers and non-alcohol drinkers, the same distribution of smokers and non-smokers, the same mix of fitness levels between individuals, and so on. In essence, randomized control trials eliminate bias and minimize confounding factors. Most importantly, randomized control trials can establish causality (although even these types of trials aren't watertight as it is difficult to make only one change in a nutrition experiment).

CONFUSION AND CONFOUNDING

When hypotheses from nutrition studies are put to the test in controlled trials, the association often turns out to be just that: an association and not a cause. For example, when the 'fruit and vegetable consumption is associated with a lower risk of heart disease' hypothesis is put to the test in controlled trials, we find that fruit and vegetables don't cause a lower risk of disease. Some controlled trials on fruit and vegetables even show completely unexpected results: an *increase* in disease risk. But we rarely hear when supposedly established theories are refuted in controlled trials. Most controlled studies show that it isn't the fibre in fruit and vegetables, nor is it the vitamins and antioxidants that make them appear healthy. So, what exactly is it about fruit and vegetables that make them look good in epidemiological studies?

It could be confounding factors. This picture below of a burger, chips and fizzy drink encapsulates the problems with confounding factors. The headline that might go with this picture could be, 'Meat causes cancer'. But how exactly do the researchers know that it was the meat in the burger that caused cancer? It could have been the bread, the chips, the fizzy drink or something that wasn't even in the picture (perhaps people who eat burgers are more likely to smoke?). Even though the authors of studies say that they have accounted for confounding factors, they often acknowledge that it is difficult to know about all these factors. The image below summarizes how confounding factors and statistics in certain types of nutrition studies can be misleading.

Is it the meat in the burger that causes an association with cancer?

Or is it the nitrates in the bacon?

Or is it the sugar in the fizzy drink?

Or maybe the gluten in the bread?

Or the trans fats in the cooking oil?

Or the carbs in the chips?

Or maybe the person who ate this meal didn't exercise regularly?

MCDONALD'S VS MCDONALD'S

While on the subject of fast food, one of the most famous, or infamous, 'investigations' of the impact of fast food was carried out by Morgan Spurlock for his documentary film *Super Size Me*. The premise of Spurlock's experiment was that he would eat exclusively at McDonald's restaurants three times a day, and not eat or drink anything anywhere else for thirty days. He would choose the standard meals from the McDonald's menu, but if he was offered the option to 'super size' the meal – a larger portion of fries and a larger fizzy drink – he would accept. By the end of his experiment, Spurlock had gained 11kg (24 pounds), going from 84kg (185 pounds) to 95kg (209 pounds), and he also claimed to suffer depression and heart palpitations during the thirty days.

Now consider the example of Kelly Hogan, a music teacher from North Carolina. In 2005 she was obese and weighed in at 119kg (262 pounds). After trying many diets and failing to lose weight, she eventually took drastic action, and as a result lost nearly 59kg (130 pounds). How? By sticking to ... a McDonald's diet.

Like Spurlock, Hogan also dined at McDonald's, but with an important difference – she ate only the meat patties. She ate on average eight quarter-pound meat patties per day, which earned her nickname of 'Quarter' at her local McDonald's. Hogan discarded the buns and avoided fries and sugar-laden drinks. After her experiment ended, she maintained her new weight by eating meat, eggs and cheese. When it came to Kelly Hogan's obesity, carbohydrates were the culprit, and it's not just McDonald's that offers those.

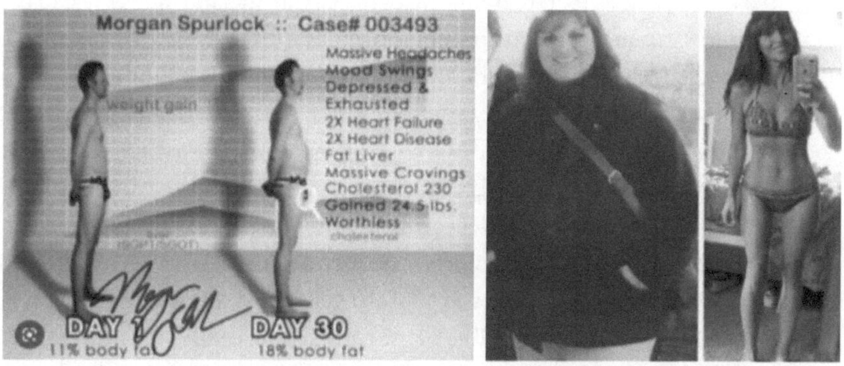

Before and after pictures of Morgan Spurlock and Kelly Hogan.

A MAN OF HIS TIMES

Some say that it was the result of the bias of one particular researcher, Ancel Keys, that led to the current flawed nutrition guidelines. A humble biologist, albeit one who made it to the front cover of *Time* magazine in 1961, Ancel Keys was convinced that dietary fat was the cause of heart attacks. Keys came to this conclusion after becoming aware that the incidence of heart attacks had been rising dramatically since the Second World War and that businessmen appeared to be at the highest risk. He conducted a study on 283 businessmen in Minneapolis and found that they were smokers, had high blood pressure and high cholesterol levels. Keys homed in on cholesterol and went on to make it his life's mission to show that eating fat results in high blood cholesterol levels and causes heart attacks. In fact, we now realize that it is another fat in the blood, known as 'triglyceride', that is behind heart attacks. And the foods that make our triglyceride levels go up the most are not fatty foods, but carbohydrate-rich foods.

To test his hypothesis, Keys launched the "Seven Countries Study". The main purpose of this well-known study was to find a relationship between diet and cardiovascular disease in middle-aged men. What the study found was that there was, indeed, a strong correlation between saturated fat consumption and heart disease. Though, as with all observational studies, they tell us diddly squat about cause and effect. There have been many criticisms of this study; firstly the selection of the countries has been challenged. Keys' critics contend that only countries that fit in with his hypothesis were selected. Besides this "selection bias" there were other important confounding factors. For example, the study was done during the period of lent in Crete when the consumption of meat, eggs and cheese is forbidden. Since these foods are the main sources of saturated fat, dietary data collected during lent is not representative of the Cretan diet during the rest of the year. In fact, saturated fat consumption normally halves during lent in Crete. Prominent scientists of the day criticized the Seven Countries Study, disagreed with Keys' hypothesis and strongly advised against recommending a low-fat diet to the American public. Nevertheless, a grand low-fat, high-carbohydrate diet experiment on the Western world was put into motion, on the basis of highly questionable epidemiological evidence.

Then there is the question of why the results of a double-blind randomized controlled trial (the absolute pinnacle of trials) conducted between 1968 and 1973 were buried for several decades. The Minnesota Coronary Experiment, as it is known, was a large and rigorous dietary trial conducted with the aim of assessing the effects of lowering cholesterol on heart disease. The 9,000 men and women in this trial were divided into two groups. Both groups were fed a typical American diet containing equal amounts of fat. However, one group's diet was high in saturated fat and other group's diet contained half that amount of saturated fat. To replace the missing half of the saturated fat, the test group were fed vegetable oil instead (in the form of corn oil and corn oil polyunsaturated margarine). The study was set up by the researchers of the Seven Countries Study – Dr Ivan Frantz and Ancel Keys – who wanted to show that high blood cholesterol causes heart disease.

What made this trial better than modern-day nutrition studies? The fact that all the study participants ate only what was supplied, didn't have access to any other food at all and the researchers didn't have to rely on food diaries. How was all this guaranteed? Because it was carried out in mental institutions where the patients had no choice regarding their food. Besides being the perfect type of trial, this study was a rare example of a diet study in which there was only one change: the type of fat was different in both arms of the study; all other foods and calories were exactly the same. So, what did this study show?

As Ancel Keys and Dr Ivan Frantz had expected, a year of eating vegetable oil did make cholesterol levels fall. The big surprise, though, was that the *lower* the cholesterol level, the *higher* the risk of death. Autopsies on the participants showed that 41% of the vegetable oil group had a heart attack, whereas only 22% of the high saturated-fat group had a heart attack. It isn't known why the full results of this study, which pitted saturated fat against corn oil, weren't published at the time (the results were partly published in 1989). It took nearly half a century for the full results of the Minnesota Coronary Experiment study to get published, and then only because a prominent researcher asked Dr Robert Frantz, a son of Ivan Frantz, whether he knew where the original records were. Dr Robert Frantz drove ninety minutes to his childhood home to look for the records. On the third trip home he unearthed yellowed documents and old computer tapes in dusty, unlabelled boxes in the basement. The tapes were translated into formats that modern computers can read and the data was analysed using up-to-date statistical methods. Could it be that Keys and Frantz didn't publish the full results of their study because they were reluctant to believe results that conflicted with their 'cholesterol causes heart attacks' theory?

Nowadays it is rare to find a study as meticulous as the Minnesota one. Firstly, modern ethics committees wouldn't approve of doing dietary studies on patients in mental institutions. In order to conduct a study such as this outside of an institution, one would have to confine participants in a laboratory and supply their meals for a whole year. Besides not being practically feasible, it would be prohibitively expensive. Just another example of the reliance, in spite of the flaws, on cheaper alternatives – epidemiological studies and food diaries.

SUMMARY

Let's repeat the question posed at the beginning: would we be happy to take a drug that had only been tested by the pharmaceutical company through an epidemiological-type study where the efficacy of the treatment was judged by relying on volunteers self-reporting on the dosages taken, and where there was little or no control of confounding factors designed into the study? Or would we prefer that the drug had been subjected to a randomized control trial, one that could provide overwhelming evidence of cause and effect, and whether that effect was beneficial or otherwise? We'd expect the pharmaceutical company to use the highest possible standards to test something we are going to ingest into our bodies. And yet many of us don't think twice about making long-term changes to our diet based on epidemiological studies reported in the media.

So how is this book different to any other diet book? Because this book focuses on hard objective evidence. We place less weight on evidence from epidemiological diet studies and place greater weight on evidence from randomized control studies. We favour evidence from modern-day paleopathology, the science concerned with diseases found in ancient human remains. And we prefer evidence from our understanding of how our ancestors evolved over millions of years. Indeed, it's tempting to characterize evolution as the ultimate 'randomized study', one where chance fluctuations in our genes results in survival of the fittest, and extinction of the rest.

What does the evolutionary and palaeopathology evidence show? It shows that the health of *Homo sapiens* declined after the widespread introduction of agriculture and change to eating foods that we were not adapted to eat (we'll cover this more fully in chapter 14). Rather, we successfully evolved, thrived and were healthy for hundreds of thousands of years on an animal-based diet.

Unfortunately, we have current dietary guidelines that are at odds with evolution. So at odds, in fact, they are the complete opposite of our evolutionary diet, as we shall now explore.

Section Two

Monumental Folly

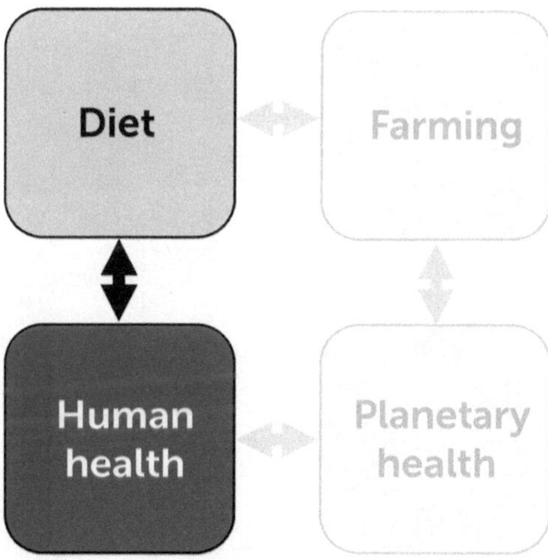

CHAPTER 4

MONUMENTAL FOLLY

The food pyramid

'The proper order of things is often a mystery to me.'
Lewis Carroll, *Alice's Adventures in Wonderland*

IT'S NOT USUALLY WISE TO RELY ON American cartoons as sources of dietary advice. *Popeye the Sailor*, who first appeared in 1929, attributed his super-strength to eating iron-rich spinach; Popeye apparently inspired Americans to increase their spinach consumption by a third. Trouble is, spinach isn't actually a good source of iron. To begin with, the amount of iron that was believed at the time to be contained in spinach had been overestimated by a factor of ten, owing to the misplacement of the decimal point in the figure. This error was eventually corrected in 1937, with the decimal point restored to its rightful position (that is, one place further to the left). And then, the iron that spinach offers is the wrong sort. Plants, unlike meat, do not contain haem iron, which is the type of iron easily absorbed by the human gut. The iron found in plants is not easily absorbed – it's not *bioavailable*. Whereas the haem iron in animal foods is very bioavailable. But the myth that spinach is a good source of iron persists.

Which brings us to *South Park*. In the episode 'Gluten Free Ebola', the character Eric Cartman averts dinner-time disaster on the East Coast of the US. A USDA dietician is explaining that their food pyramid was built to show how much people should eat of the four basic food groups. Cartman intervenes: the pyramid, he warns, is 'upside down'.

But what is the food pyramid? What did Cartman mean by 'it's upside down'?

THE FOOD PYRAMID

The food pyramid was originally developed in Sweden in 1974. The United States Department of Agriculture then adopted this idea of presenting diet recommendations in the form of a pyramid in 1992.

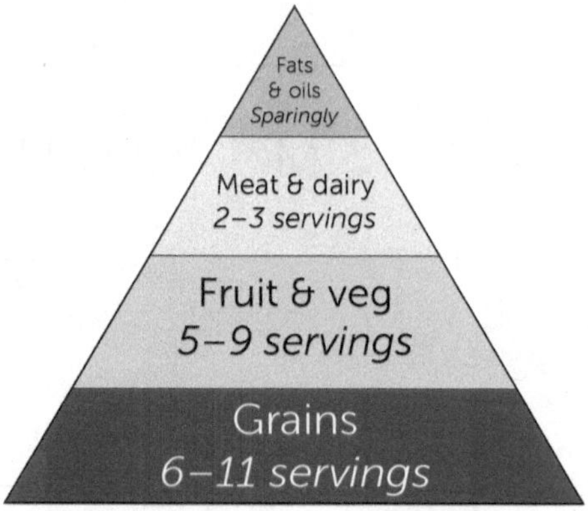

A simple form of the food pyramid, with USDA-recommended daily serving amounts.

As you can see, the pyramid was heavily biased towards plant foods. Of some twenty or so daily servings, 80% or more were meant to come from grains, fruit and vegetables. This prejudice reflected the vilification of meat and other animal foods, as exemplified by Ancel Keys' false hypothesis, from the 1950s onwards, that saturated fat, as found in animal fats, was a cause, if not the prime cause, of coronary heart disease. This hypothesis, though roundly contested by many doctors and dieticians at the time (and since), was enough to convince George McGovern's Senate Select Committee in the 1970s to introduce guidelines that Americans should eat less fat and instead eat more carbohydrates.

THE STANDARD AMERICAN DIET

The Standard American Diet (commonly termed the 'SAD diet') is the diet that is actually eaten in America and most developed countries. As we've seen, starting fifty years ago, Americans were advised by various government agencies to reduce cholesterol and fat, especially animal fat, and increase carbohydrate intake in their diets. In many respects, the American public has been compliant with their government guidelines on nutrition. For example, below are some of the official recommendations in the USA and what the Americans actually ate over the last fifteen years:

	Recommendations (percentage of total calories)	**Actually eaten**
Carbohydrate	45% to 65%	50%
Fat	20% to 35%	33%
Protein	10% to 15%	15%
Cholesterol	300mg/day	276mg/day

As you can see, the Americans did as they were told (there's always a first). However, if we look at a graph of obesity rates over the last fifty years, it appears that the introduction of the low-fat guidelines coincides with the beginning of the obesity epidemic.

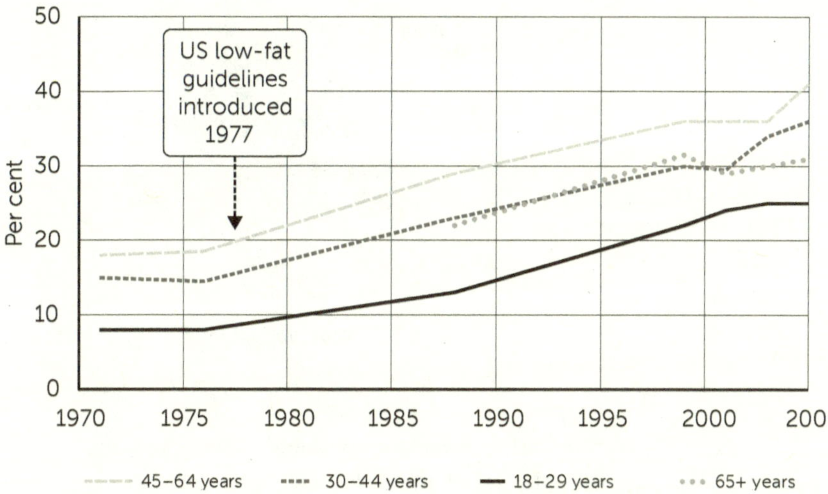

There are many who are of the opinion that the low-fat nutrition guidelines have directly or indirectly led to the current obesity epidemic. But, as association does not mean causation, this graph could never prove that the nutrition guidelines caused the obesity epidemic, despite many large recent studies supporting the idea that a low-fat diet is associated with obesity and an increased mortality risk. Alas, many of these recent studies were also epidemiological studies. Ideally, the hypothesis that 'a low-fat and high-carbohydrate diet causes diabetes, obesity and an increased risk of mortality' should have been tested in randomized controlled trials. Instead, many countries have been the guinea pigs of a dubious and untested hypothesis.

THE EATWELL PLATE

Over time the pyramid was replaced by a plate. In 1994 'The Balance of Good Health', a picture of food divided into groups around a plate, was launched by the UK Department of Health. It evolved, and by 2016 became the Eatwell Guide. As described at the time by Zoë Harcombe (researcher, author, blogger and public speaker in the field of diet and health), the Eatwell Guide was 'formulated by a group appointed by Public Health England, consisting primarily of members of the food and drink industry rather than independent experts'. Which, if true, sounds like a recipe for whipping up conflicts of interest. The plate is how the UK and US dietary guidelines are now usually presented.

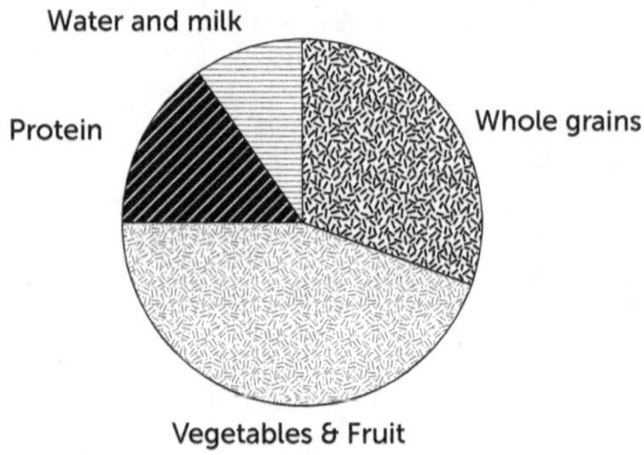

The Eatwell Plate.

Though the diagram had evolved from pyramid to pie chart, the proportions of plant and animal foods hadn't changed, with around 75% of the recommended daily intake as shown on the plate coming from grains, fruits and vegetables.

But whether presented as a pyramid or a plate, the current food guidelines are a monumental folly, a folly resulting from the sort of flawed dietary studies that we covered in the previous chapter. Modern dietary advice restricts animal fats and substitutes them with carbs. The advice also spurns high-quality, vitamin- and nutrient-rich meat and offal and replaces them with low-quality fruit and vegetables, where nutrients either have poor bioavailability or are completely missing. And guidelines encourage the use of vegetable oils, which contain the wrong mix of fats needed by the brain and body.

Modern dietary advice as summarized by the plate or pyramid couldn't be more wrong.

The diet deception has similarities with the Post Office scandal in the UK of the first two decades of the twenty-first century (and continuing as we write). Here was a once trusted national institution that insisted that its prosecution of sub-postmasters was justified, in part because it believed the evidence that supported the prosecutions was 'robust'. Compare that with the advice on diet given by Public Health England and the NHS, backed up, they would doubtless claim, by expert advice and 'robust' studies. But here's where the similarity falls down – the diet deception is much, much bigger.

Rather than the thousand or so sub-postmasters who were persecuted on the basis of poor or suppressed evidence, the dietary guidelines, obsession with plant foods is ruining the lives of millions. With justification, Dr George Mann, associate director of the Framingham Heart Study, said that '[saturated] fats and cholesterol in the diet were not the cause of coronary heart disease. That myth is the greatest deception of the [twentieth] century, perhaps of any century.'

One of the authors recently complained to the BBC when it repeated the mantra that saturated fats were linked to heart disease. The article said that saturated fat 'raises blood cholesterol and raises the risk of heart disease.' The author attached many references and studies that cast doubt on the claim. Their response? The 'details provided … reflect [sic] the information given by both the National Health Service and the British Heart Foundation … (please note that the BBC is not responsible for the content of external websites)'. It seemed they hadn't paid any attention to the references provided. Complacently, they acted as a mouthpiece for the NHS, with no scrutiny, no challenge, no balance, no objectivity. Unable, or not interested, to properly weigh and consider contrary opinions, this mindset is typical of the BBC's, and the wider media's, approach to diet reporting.

Then look at this statement. 'In particular, I want to say sorry not just for the actions which led to life-altering and life-limiting illness, but also for the failures to clearly communicate, investigate and mitigate risks to patients … [and] for a collective lack of openness and willingness to listen, that denied patients and families the answers and support they needed.'

This is the chief executive of the NHS in 2024, responding to the publication of the Infected Blood Inquiry Report. The authors believe a similar lack of openness and willingness to listen extends to those in the NHS, Public Health England and government responsible for agreeing and issuing our current dietary guidelines.

NUTRITION 101

To better understand the next few chapters of the book it is worth explaining a few facts of life. Not the birds and the bees – rather, metabolism and food. Metabolism is about converting food into what we need. In order to maintain life, food, water and oxygen provide all the substances needed for metabolism. The aim of every species is to survive, thrive and reproduce (so, okay, birds and bees do come into it) and, to do this, we have to metabolize food. But not just any food; we need food that contains certain nutrients.

There are two main types of nutrients – 'macronutrients' and 'micronutrients'. Macronutrients provide energy and can be eaten in large amounts; they consist of fats, proteins and carbohydrates. Of the three macronutrients, proteins and fats are needed for energy, for growth, to repair our bodies and for other important functions. Carbohydrates, on the other hand, only provide energy and aren't needed for any vital processes.

Micronutrients, as the name implies, are needed by our bodies only in small amounts and consist of vitamins, minerals and certain essential fats. Vitamins and minerals do not provide any energy; instead they act as catalysts for the thousands of chemical reactions that are constantly going on in our bodies, reactions that would otherwise grind to a halt.

WHY ALL CARBOHYDRATES END UP AS SUGAR IN OUR BODIES

Before looking at how we get energy out of the carbohydrates, proteins and fats in food, we will briefly explain what carbohydrates and fibre are. As far as nutrition goes, all carbohydrates are either sugars, starch or fibre. Then there are digestible and non-digestible carbohydrates; sugars and starches are digestible, whereas all types of fibre are non-digestible. To complicate matters a bit more, fibre can be soluble or insoluble.

Cellulose is one of the key components of insoluble fibre and pectin is the main type of soluble fibre (both are found in the cell walls of plants). The key point here, though, is that all types of fibre are indigestible, i.e. the human gut is unable to break down and absorb fibre in any form. Paper, cotton, wood are all fibrous materials composed of cellulose – just as we don't need to eat these, so we will explain in chapter 6 why we don't need to eat any fibre at all.

To come back to digestible carbohydrates, these can be either 'simple carbohydrates' (sugars such as glucose, fructose and galactose) or 'complex

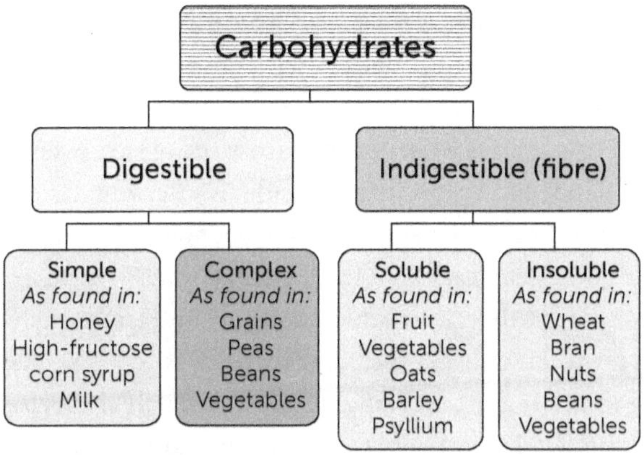

The different types of carbohydrates in food.

carbohydrates' (which are all starches). The thing is, all complex carbohydrates are made up of simple carbohydrates (sugar molecules) strung together in various configurations. And *all* complex carbohydrates, whether they are from whole grains, beans, vegetables or any other type of food, are ultimately turned into sugar (simple carbohydrates) by our bodies.

The figure above shows all the different types of carbohydrate in an easy-to-understand flow chart.

From here onwards, when we refer to 'carbohydrates', we are speaking of sugars or starches only (and not fibre) unless stated otherwise.

HOW WE GET ENERGY OUT OF CARBOHYDRATES, PROTEINS AND FATS IN FOOD

First, these macronutrients are broken down into other intermediates. Proteins are broken down into amino acids, carbohydrates are broken down into glucose, and fats are broken down into fatty acids or converted to ketones (among other substances). Then these various intermediates are converted into 'adenosine triphosphate (ATP)', which is the form of energy that is ultimately used by cells (ATP is also known as the 'universal energy currency' for all life forms).

When it comes to storing energy from food, excess glucose can be stored as glycogen in the liver and muscles, but only in fairly limited amounts. Once glycogen stores are full, any excess energy is stored as fat throughout the body. Unlike glycogen, our ability to store fat is enormous (as is obvious when we look at an obese person). As a result, glycogen stores, being limited, can only provide energy for a day at the most, whereas fat reserves can provide energy for several weeks.

What is arguably even more important to know about nutrients is whether they are 'essential' or 'non-essential'. To be lacking in non-essential nutrients isn't a problem because our bodies can make them from other nutrients. For example, some amino acids (which are what proteins are made of) are essential, while others are not. If we happen to be low in a particular non-essential amino acid, our bodies can convert other amino acids into the one that is lacking. But the amino acids that our bodies can't make – the 'essential' ones – have to be provided by our diet. Similarly, there are other essential nutrients, such as vitamin B12 and iron, that we can only get from our diet.

At this point, owing to the importance of glucose, it is worth clarifying a common misconception about it. Many people are of the opinion that carbohydrates are essential nutrients because they are the main source of glucose for those on a standard Western diet. However, even though glucose is an *essential molecule* in the human body, it is not an *essential nutrient* in the human diet. This is because it is easily synthesized from other nutrients. Therefore, neither carbohydrates nor glucose are essential components of our diet.

The consequences of a lack of actual essential nutrients are wide-ranging, from mild fatigue to much more serious symptoms, and even death. Of particular relevance to this book is the fact that there are many essential nutrients that cannot be provided by a plant-food only diet, whereas an animal-food only diet is able to supply every single nutrient that we need.

But we don't eat nutrients – we eat food. The physical structure of food and the way the molecules within a food interact affect how we digest and absorb the nutrients it contains. This is sometimes known as the 'food matrix'; food is much more than just a list of nutrients. And then there are 'anti-nutrients' that need to be taken into consideration – plant foods are packed with anti-nutrients (animal foods have virtually none). These anti-nutrients in plant foods inhibit the absorption of other nutrients. Moreover, nutrition science is a relatively new discipline, and every few years we discover a new essential nutrient, such as choline and vitamin K2 in the last couple of decades.

LET'S GO

Despite, or indeed because of, its faults, we'll use the food pyramid in the following chapters as a framework. It will help us explore the inadequacies of modern dietary guidelines, such as Eatwell, and to explain how an animal-based diet like the Sapiens diet corrects these shortcomings. These shortcomings have contributed to a tripling of obesity rates, a tripling of the incidence of type 2 diabetes, as well as poor brain health, both in terms of its physical structure and in mental well-being.

Over the course of the next few chapters, we'll deconstruct the pyramid from the bottom upwards. Then we'll rebuild it, finishing with something that would merit Cartman's approval. Still a pyramid, but now the right foods, in the right order and in the right amounts.

We'll proceed as follows:

CEREAL KILLERS
Why it's disastrous to remove animal fats as a source of fuel for the body and replace them with glucose from starchy and sugary foods.

FIBRE – A NATURAL JUNK FOOD
Why fibre isn't required in our diet.

VITAMIN C AND THE FRUITLESS DIET
Why fruit and vegetables are not a good source of vitamins and nutrients, but animal foods such as liver are.

FREE THE RADICALS
Why we don't need antioxidants or other supplements on an animal-based diet.

WHY VEGANS HAVE SMALLER BRAINS
The physical and mental health benefits of animal foods for the brain.

JUST STOP VEGETABLE OIL
Why seed oils are bad, and animal fats are good.

THE KEY TO EVERYTHING DIET
Why ketogenic diets are healthier than low-fat diets.

DOUBTS ABOUT GOUT
Why we don't need to worry about gout on the Sapiens diet.

MIND THE GAP
The importance of a leaky gut.

MEAT AND NO VEG
With the pyramid inverted, we bring together why an animal-based diet is the healthiest diet.

Let's go to the base of the pyramid and begin with 'cereal killers'.

CHAPTER 5

CEREAL KILLERS

Carbs are the 'new smoking'

'We now eat in two weeks the amount of sugar our ancestors of two hundred years ago ate in a whole year.'
John Yudkin

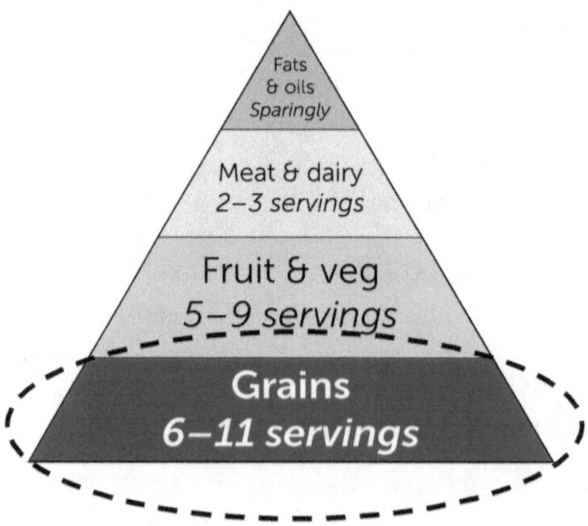

'Cereal killers'. Or perhaps, 'killer carbs'. Provocative phrases, but this chapter aims to show that there is justification for them, and justification to ditch cereals and other carbohydrates altogether from our diet. The UK's nutrition guidelines endorse meals based around 'potatoes, bread, rice, pasta or other starchy carbohydrate'. The 'grains' at the base of the original food pyramid, and the 'cereals' in the title of this chapter, include the usual suspects: wheat, barley, rye, oats, rice, sorghum and millet.

This chapter is about cereals, carbohydrates, obesity, diabetes and insulin. Why devote a whole chapter to these topics? Because obesity and diabetes are massive problems. Since the mid-1990s, the number of people in the UK diagnosed with diabetes has more than trebled, from 1.4 million to 3.9 million. This figure is expected to continue to rise to well over 5 million people in the next few years. Looking to the international figures, 9.3% of all adults in the world have full-blown diabetes. And as far as obesity goes, 26% of adults in England are now obese and a further 38% are overweight, almost double what it was in the 1970s.

Diabetes, obesity and high insulin levels are all associated with an increased risk of heart attacks, strokes, cancer, kidney failure, blindness, lower limb amputation, fatty liver, gout, impotence and acne. And globally, heart attacks, strokes and cancer account for more than half of all deaths.

So, yes, cereal, carbohydrates, obesity, diabetes and insulin matter a lot.

Yet it has not escaped notice that since the promotion of high-carb diets that diabetes and obesity have escalated. Why are carbs to blame? Because diabetes and obesity are associated with overconsumption of sugar. Sugar is a common ingredient added to many things that we consume. What many people don't realize is that *all carbohydrates turn into sugar in the human body*, no matter whether they are refined carbohydrates or unrefined carbohydrates. Whether it is sliced white bread, artisan sourdough, a wholemeal cob, white rice or brown rice, white potato or sweet potato, they all end up as sugar in our bodies.

Association does not prove cause, and we cannot say for certain that the nutrition guidelines are responsible for increasing diabetes and obesity in the population. But has it achieved anything in tackling these diet-related health issues? It is surely fair to say that the guidelines have failed to curb or make any dent in the levels of diabetes and obesity in the UK.

In the next few pages we'll take a closer look at diabetes, and the role insulin plays in weight gain and obesity. And we'll address the importance of different diets, either as the cause of diabetes and weight gain, or as a preventative or treatment for those conditions.

To begin with, we need to introduce carbohydrates.

WHAT ARE CARBOHYDRATES?

Although covered in 'Nutrition 101', we will do a brief recap of carbohydrates. Carbohydrates are the sugars, starches and fibre found in a wide variety of foods. In this chapter we will cover sugars and starches – fibre gets its own chapter. Nearly all carbohydrates come from plant foods, while animal foods contain virtually none. Carbohydrates are sometimes divided into 'simple carbohydrates' and 'complex carbohydrates'. Simple carbohydrates are sugars, and complex carbohydrates are starches. Sugars occur naturally in fruits, vegetables and honey (as sucrose, glucose and fructose) and in milk (as lactose). Sucrose (table sugar) is the sugar that is usually used in jams, biscuits, confectionery and cakes. Starch is abundant in cereals such as wheat, maize and rice, in starchy vegetables like potatoes, and in grain-based foods such as bread, pizza or pasta.

What isn't widely known, though, is that the starches in all these foods are made of sugars (such as glucose). Starches are essentially chains of glucose bound together in various amounts into much larger units. Starches (complex carbohydrates) are digested more slowly and release glucose into the blood more gradually than sugars (simple carbohydrates) do. The bottom line is that all complex and all simple carbohydrates end up as sugar in our bodies – the main difference between them is the speed with which this occurs.

Carbohydrates are also one of the three kinds of macronutrients that our bodies can use to provide us with fuel for energy. Fats and proteins are the other two. We measure the amount of energy they provide in calories. Our bodies don't mind where our calories come from. If we reduce our calories from carbohydrates, by eating less bread perhaps, then we can fill the gap with calories from fat or protein, or both, by eating meat for example.

All carbohydrates are ultimately broken down into sugar during digestion. Glucose is either used as a fuel to provide us with energy, or if we have a surplus of calories, glucose gets converted to glycogen or fat. However, even though some glucose is essential for our brain to function properly, carbohydrates are not essential to provide glucose – our bodies can make all the glucose we need from protein.

But energy isn't everything. We also need nutrients from food to build and maintain the cells and tissues of our body. Proteins and fats are essential in our diet to provide these body-building nutrients. In fact, we can get *all* the calories and body-maintaining nutrients we need from animal foods. It is impossible, though, to get all the nutrients that we need from a diet that is entirely plant-based. The bottom line is that we can be completely healthy, if not healthier, without eating any carbohydrates, or any plant foods. Or to put it another way, plant foods are not an essential component of the human diet, whereas animal foods are. And we will see in chapter 7 that we can easily get every single vitamin and mineral that we need from animal foods.

What happens when we do eat carbohydrates? Digestion of carbohydrates begins with saliva in the mouth when we chew our food, but the majority of digestion occurs in the small intestine. Here, an enzyme breaks down complex carbohydrates (such as those in bread) into simple sugars, such as glucose, which are then absorbed into the bloodstream. In the average healthy adult there isn't very much glucose circulating in the blood at any one time (roughly 5g, which is the equivalent of a teaspoon of sugar). Very low or very high blood glucose levels are extremely dangerous, so our bodies have a highly sophisticated control system to regulate the level. And this is where we come to insulin.

INSULIN AND DIABETES

But what exactly is this insulin? Some call insulin the 'master' hormone. Others dub it the 'fat storage hormone'. Insulin helps your body turn food into energy and manages your blood glucose levels. The more blood glucose levels go up, the more insulin goes up to lower it. This critical hormone is secreted by the pancreas and determines how much glucose gets into and out of our cells.

WHERE DOES GLUCOSE COME FROM?

Carbohydrates in our food are broken down into glucose within one and a half hours after being eaten. Protein and fat also come into the equation, as these can be sources of glucose. Depending on how much is consumed and what the body needs, roughly half of any protein in our food is converted to glucose, but only very small amounts of fat consumed (10% or less) are converted to glucose within three to four hours of eating. Because glucose makes insulin go up, this means that carbohydrate-containing foods make insulin levels go up much more than would protein or fat. Wholewheat pasta, wholemeal chapattis, oatmeal and porridge may sound healthy, but they are all just glucose in disguise.

In diabetes, blood glucose levels become abnormally high. You have probably heard that there are two types of diabetes. Type 1 diabetes is an autoimmune condition in which the body is unable to produce enough insulin to control blood glucose levels. But our main concern here is type 2 diabetes, which begins when we produce abnormally high amounts of insulin, as a result of lifestyle factors. (From here onwards when we refer to 'diabetes' we are talking about type 2 diabetes unless stated otherwise.) Genetic factors do come into it though, – we all have a different susceptibility to diabetes – so some can get away with

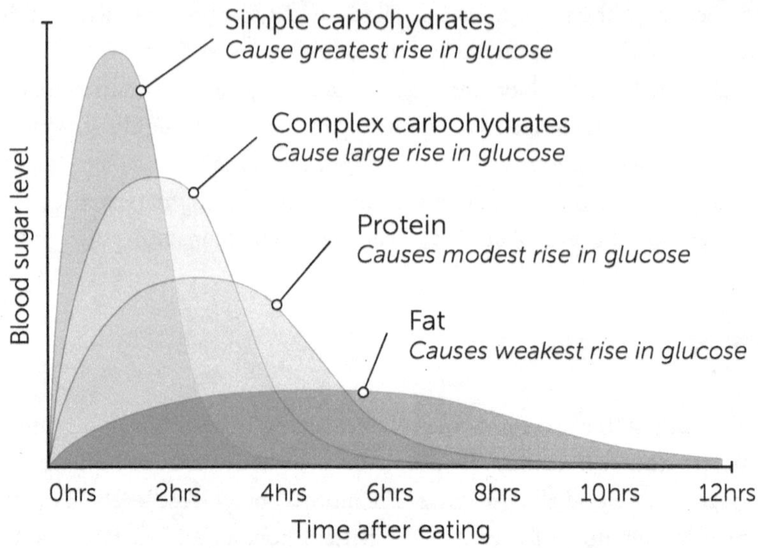

Effects of carbohydrates, protein and fat on blood glucose levels (source: adapted from *Gestational Diabetes UK*).

eating large amounts of carbohydrates and others can't. Sadly, those who don't get away with it, end up with diabetes. You can't change your genes very easily (yet), but you can change your lifestyle. Of the lifestyle factors, diet is paramount, but more on that later.

An excessively high blood glucose level is like poison to the cells in our body, while an excessively low blood glucose level can be fatal. There is a sweet spot in the middle. When we eat foods that contain carbohydrates, our blood glucose increases; the more carbohydrates we eat, the more our glucose goes up. The more glucose goes up, the more insulin we secrete to bring glucose back down. Conversely, eating fewer carbohydrates means we produce less insulin. Which is why what we eat is so important. Most people know that insulin has something to do with controlling glucose, but not many know that it is the *only* hormone that lowers blood glucose. This is important, because one of the key issues at the heart of most Western diseases is an abnormally high insulin level.

THE DIABETES CONTINUUM

STAGE 1: HIGH INSULIN LEVELS

When we eat a meal containing carbohydrates, our blood glucose level goes up and our insulin level also rises to control this. If we are not diabetic, glucose and insulin return to normal levels within two hours. But regularly eating a diet that is high in carbohydrate results in chronically high insulin levels. This is the first step on the path to diabetes. In this first stage insulin levels are very high but glucose levels are normal.

STAGE 2: PRE-DIABETES AND INSULIN RESISTANCE

In this second stage of the road to diabetes, insulin levels remain extremely high, but in spite of this, blood glucose levels start climbing, either to the top end of the normal range, or occasionally even beyond. The body at this stage has developed 'insulin resistance' – cells are becoming 'deaf' to insulin telling them to take glucose out of the blood. To make up for this, the pancreas has to 'shout' by secreting ever more insulin to keep blood glucose levels normal.

STAGE 3: TYPE 2 DIABETES

In this final stage, despite massive amounts of insulin being pushed out by the pancreas, normal blood glucose levels cannot be maintained and we end up with full-blown diabetes. It can take decades for the first stage of high insulin levels to progress to full-blown diabetes. If high insulin levels and diabetes continue for many years, the pancreas eventually burns out. At this stage, insulin production plummets and type 2 diabetes becomes like type 1 diabetes.

However, let's not forget that even though excessive amounts of insulin are detrimental, we do always need some insulin. In fact, not having enough insulin (as in type 1 diabetes) is an urgent and life-threatening condition. Excessively high insulin levels, though, go in for a slow kill. What many don't know (including many doctors) is that most of the diseases that are linked to diabetes are also linked to high insulin levels. The blood test that is normally used to check insulin levels is a simple fasting insulin test. But this can be a misleading test. It is possible for fasting insulin levels to be completely normal, even though insulin levels after meals may be astronomically high. And it is excessively high insulin levels after meals that cause problems.

Currently, the best way of assessing insulin levels after a meal is to drink a standard glucose drink and then test the blood insulin level every thirty minutes for the next two hours. This prolonged insulin testing was pioneered by Dr Joseph R. Kraft, a very astute doctor in Chicago. He carried out this extended test on 14,000 people from the general public over the course of thirty years. His major discovery was that 80% of those tested had extremely high insulin levels, despite most having completely normal blood glucose levels. This statistic still rings true today; four-fifths of us produce abnormally high insulin levels after eating a meal – yet this Kraft test is still not routinely offered to patients. The greatest value of the Kraft insulin test is that it is an early warning; it shows that we are on the diabetes continuum many years before a blood glucose test starts becoming abnormal.

The amount of insulin needed to keep glucose levels normal can differ tenfold between individuals. I can attest to this by showing the differences between my own insulin levels and those of my brother while I was on the Sapiens diet and my brother was on a 'normal' carbohydrate diet. The graph below shows both our insulin levels during a Kraft test. At baseline, both of us had normal fasting insulin levels. However, after the 75g glucose drink my brother's insulin level rose more than ten times higher than mine. Yet his blood glucose and fasting insulin have always been plumb normal (which is why he kept questioning the need to do this Kraft test – his main objection being that he had to be jabbed with a needle five times in the space of two hours).

Insulin levels after a glucose drink.

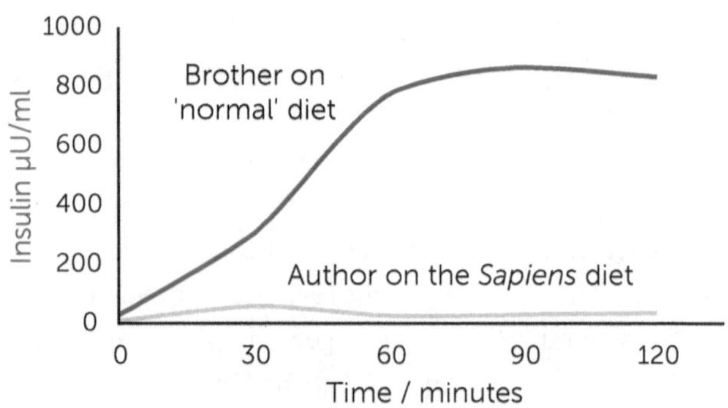

Because a diagnosis of diabetes depends on glucose, many don't know that their insulin levels might be abnormal and are lulled into a false sense of security. Plus most medical practitioners are not aware of the Kraft test. Measuring glucose to diagnose diabetes is the norm, even though the Kraft test can predict diabetes many decades earlier. Rather than cholesterol or glucose blood tests, it is this Kraft insulin test that we should all be doing.

MEAT ISN'T THE 'NEW SMOKING', CARBS ARE

In September 2022, a piece appeared in *The Times* with the title 'Eating red meat should be as vilified as smoking'. Here's an extract from it:

> We need to build a taboo around red meat, as we once did around smoking. Diners tucking happily into a rare steak should be treated with the same kind of cold outrage as a smoker lighting up next to a child. Fancy restaurants that don't wean beef off the menu should be boycotted.

While appealing to environmental concerns (misplaced concerns, as we shall see in section 3), the article also asserted that alternatives – plant foods, presumably, or perhaps protein manufactured in labs – could be healthy and issue-free (tissue-free?) substitutes.

The article generated plenty of responses, the majority unimpressed with the sanctimonious tone and uninformed content. Here's one of the replies.

> [The author] won't want to allow facts to intervene in [their] messianic vision … [The author] chooses to ignore the fact that ruminants eat grass which itself captures CO_2 whilst they pass wind … [The author] also ignores the CO_2 burden of moving to vegetarianism – fertilisers, fossil fuel emissions for trucks/harvesters … Poor stuff.

Quite so. The evidence we present in this book demolishes arguments that eating animal-sourced foods is unhealthy. There's also no need to be ashamed about eating meat from pasture-fed livestock; we'll learn later that farming in that way is not destroying the planet; in fact it helps mitigate climate change. And eating meat has no adverse health impact on the child sitting next to you. If anything, it might have a beneficial effect. Comparing meat to smoking is idiocy.

But worse (and aside from the irony that tobacco is a plant), the article missed the opportunity to nominate an outstanding candidate for the role of the new nicotine. Carbohydrates.

The image below is similar to those you see prominently displayed on cigarette packets, warning of the perils of smoking, among which is damage to blood vessels in the body's extremities. This damage leads to loss of blood supply, giving rise to tissue death, gangrene and, ultimately, amputation.

Except this picture is not of the effects of smoking. The gangrene in this foot was caused by type 2 diabetes.

A result of type 2 diabetes.

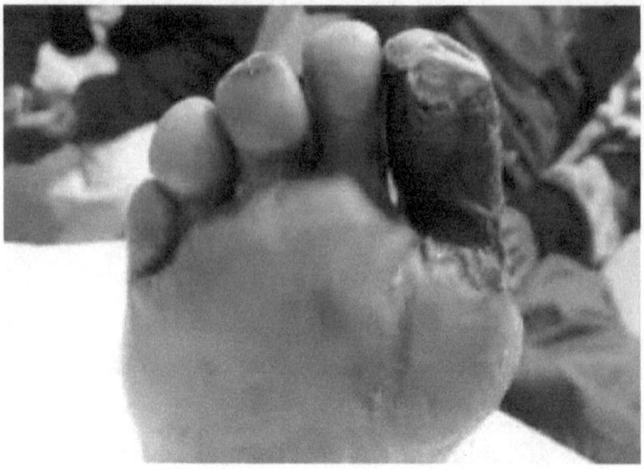

Type 2 diabetes causes damage to the lining of blood vessels, which in turn causes tissue breakdown, necrosis and gangrene. In fact, diabetes is the most common reason for limb amputations. The damage can also extend to blood vessels in the retina and kidneys, leading to sight loss and kidney failure.

Here's a thought. Restaurants – fancy or otherwise – that don't remove carbs from their menus should be cancelled. Furthermore, children should be protected from witnessing adults stuffing themselves with bread, pasta, pastries and sticky toffee pudding. Stick that in an editorial.

INSULIN – THE FAT STORAGE HORMONE

Back to biology. Once a meal has been metabolized, glucose and insulin levels drop again. When insulin goes down, another hormone, glucagon, starts to go up. Insulin and glucagon are like a see-saw – when one goes up the other goes down. The significance of glucagon is that it allows the body to start breaking down its stored energy, first from stored glycogen in the liver and muscles, and then from stored fat. To break down fat, the body needs glucagon, but glucagon can only kick in when insulin goes down. This is one of the many reasons why high levels of insulin are so problematic for our health. When we eat carbohydrates, insulin levels go up, which in turn impedes glucagon production. Without glucagon we can't break down our stored body fat. And any excess calories that are consumed are stored as fat. Consequence? We put on weight.

How much fat you gain after eating a meal depends entirely on how much insulin is released, and this depends on how much carbohydrate and protein you eat. Eating fat alone (without any carbohydrates or protein) hardly causes any insulin production. A low insulin level leaves the body able to produce glucagon and to break down stored fat. Hence eating pure fat will not make you fat. The main reason we put on weight is not how much fat we eat, but how much carbohydrate or protein we eat with the fat. You might now be wondering, 'Don't calories come into the equation?' Yes, of course calories matter. But if our bodies treated calories from fat exactly the same as they treat calories from carbohydrates, you wouldn't expect to see different weight changes when eating the same number of calories from both. In this respect, those who say that 'a calorie is a calorie' to claim that weight gain is all about how much you eat rather than what you eat are wrong. The story below exemplifies why a calorie from fat is not the same as a calorie from a carbohydrate.

AN EXAMPLE

One young chap, a certain Sam Feltham, decided to put insulin, carbohydrates and calories to the test. He subjected himself to eating a massive 5,800 calories a day on three different diets. He followed each diet for twenty-one days each and called it an 'overfeeding' study; his findings were published in a respected scientific journal. The diets tested different amounts of fat and carbohydrates (he ate the same amount of protein on all of them).

First off for Sam was a low-carbohydrate diet (6% carbohydrates and 72% fat); then followed a low-fat diet (64% carbohydrates and 23% fat) and finally a very low-fat vegan diet (68% carbohydrates and 15.5% fat). The results were as follows:

Diet	Carbs	Fat	Weight	Waist	Asthma	Other symptoms
Low carb	6%	72%	Gained 1.3kg	Lost 3cm	No asthma	Had lots of energy. Never felt hungry.
Low fat	64%	23%	Gained 7.1kg	Gained 9.25cm	Relapse of asthma on day 10	Felt lethargic. Snored at night. Didn't feel overfull.
Very low fat vegan	68%	15.5%	Gained 4.7kg	Gained 7.75cm	Relapse of asthma on day 10	Cognitively 'foggy'. Didn't feel hungry.

As you can see, it was on the low-carbohydrate diet (6%) that he gained the least weight. Compare this with the diet guidelines which tell us to eat 70% carbohydrates. Yes, we know we shouldn't read too much into studies where there is only one subject being tested. But one of the authors of this book did something similar recently. She subjected herself to a high-carbohydrate diet for four weeks. It was called 'Christmas' and so was incredibly easy. The diet included regular helpings of mince pies, roast potatoes, Christmas cake and Christmas pudding. The weight gain of 2.5kg was all in the interest of science, of course.

FROM WORSE TO BAD – CHOLESTEROL AND CARBOHYDRATES

As well as contributing to diabetes and weight gain, excess carbohydrate consumption also causes an increase in the blood level of triglycerides, which are a type of fat. It turns out that the blood triglyceride level is one of the simplest ways of finding out whether or not you have insulin resistance, and which is one of the early stages of diabetes. This may be why high levels of triglycerides are closely associated with a higher risk of heart attacks.

Consuming carbohydrates also causes a lowering of high-density cholesterol (HDL), sometimes known as 'good cholesterol'. The obsession with 'good' and 'bad' cholesterol is puzzling – whole books have been written on this subject and we will briefly cover it in the 'Good fat, bad fat' section in chapter 10. Suffice to say for now that cholesterol is needed in our bodies for many reasons. Cholesterol is used to make vitamin D and many hormones, and it is also an essential component of cell membranes. These are just a few of the reasons why our bodies need cholesterol. It is also worth pointing out here that most of the cholesterol in our bodies is made by the liver. If you cut down on cholesterol in your diet, your liver will end up making more of it.

While we are on the subject of fats and heart disease, let's take a quick look at saturated fat. We have been told time and again to cut down on saturated fat intake. Lowering saturated fat intake is one of the mainstays of current dietary advice. But what most don't know is that the saturated fat we eat has little to do with saturated fat levels in our blood. So, what does increase saturated fat in our blood? You guessed it – carbohydrates. Carbohydrate intake is, in fact, a key determinant of blood saturated fat levels. And yet in spite of all this evidence, the classic received wisdom is that fatty red meat causes heart disease.

Another reason to cut down on carbohydrates is that they increase levels of visceral fat in the body. This is important because *where* fat is matters more than how much fat you have. There are two main places where we store fat. The fat that is stored around our internal organs is known as visceral fat and this fat produces a whole host of chemicals that cause deleterious health effects, such as low-level chronic inflammation and high blood pressure. The other place where fat is stored is below our skin – this is the fat that you can see and feel – and it has been found to be protective against disease owing to the beneficial molecules that it produces. What is particularly relevant here is that a diet which is higher in carbohydrates is associated with higher levels of visceral fat accumulation, and this is one of the many harmful effects of carbohydrate consumption. However, the level of carbohydrate consumption that causes visceral fat to accumulate differs between people. Once again, a lucky few can get away with eating more carbohydrates than others can.

When everything is functioning as it should, insulin brings about the release of another hormone (leptin) which makes you feel full and stops you from eating more. When there is insulin resistance, leptin resistance usually develops too. We no longer feel full when we should, and so we carry on eating too much. If insulin resistance is corrected using a low-carbohydrate diet, the leptin resistance disappears as well. This then leads to better appetite control and makes it easier to not overeat.

CARBOHYDRATES BECOME ADDICTIVE

As we've seen, a carbohydrate-loaded meal causes the release of insulin in order to lower blood glucose levels. Owing to today's carbohydrate-heavy diets, most people are now producing abnormally high amounts of insulin. In some people, a very high insulin level causes blood glucose levels to drop precipitously. The steep drop in glucose activates other hormones that tell your brain that you

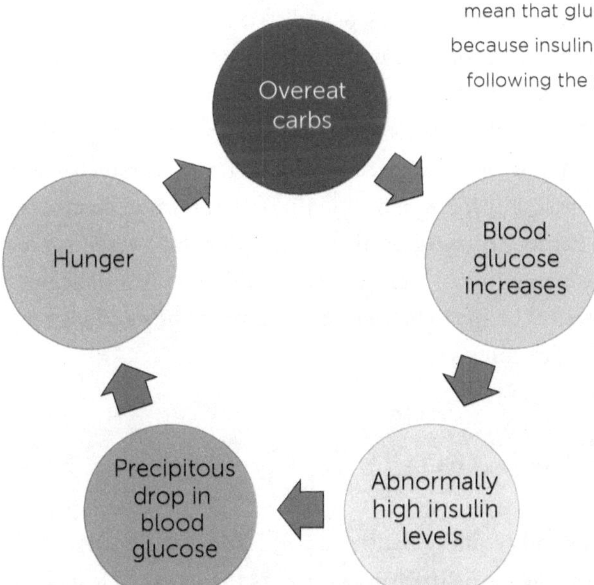

Carb addiction cycle. The effects of carb addiction and eating throughout the day mean that glucagon doesn't get a look in because insulin is always high. This is what following the nutrition guidelines does: it helps you convert excess calories to fat and stops you from burning your own body fat.

Illustration of insulin levels during the day.

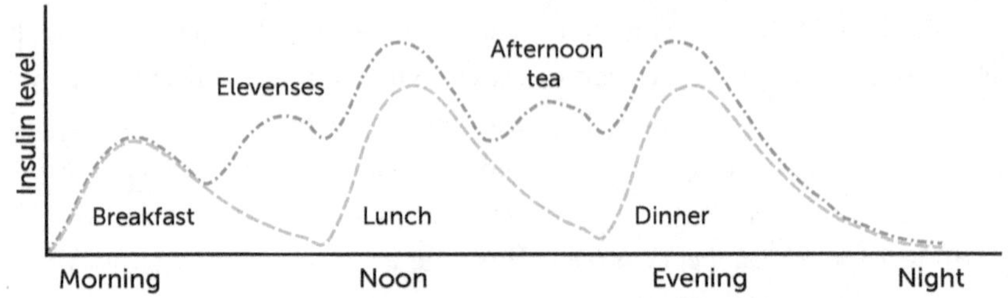

need to eat, resulting in feelings of hunger and carbohydrate cravings. This causes us to want even more carbohydrates. A vicious cycle – similar to an addiction – is created.

DRUG ADDICTIONS

I used to jokingly say that I was addicted to desserts. Little did I know that there is truth to this statement. Eating large amounts of carbohydrates stimulates the 'reward centres' in the brain and the parts of the brain that are associated with cravings – the very same parts that light up when a drug addict gets a hit. The fact that food activates similar areas in the brain as drugs do isn't new; what is new is that the type of food matters – fat is nowhere near as rewarding or addictive as carbs are.

The real killer, though, is foods that contain both fat and carbohydrate together; calorie for calorie, a combination of fat and carbohydrate gives the biggest 'hit' and much more so than foods that contain only fat or only carbohydrate. Sadly, the combination of carbohydrates with fat tastes extremely good – think cheesecake, fish and chips, pizza, toast with butter, salmon en croûte, *crème brûlée*, spaghetti carbonara, burgers, deep-fried Mars bars (then again, maybe not), chicken korma and naan – the list is as endless as it is delicious. Modern meals are the perfect recipe to help store fat on your body and prevent you from using your own fat stores.

ARE ALL CARBS CREATED EQUAL?

Are all carb-containing foods equally bad? Are the carbs in broccoli as bad as the carbs in cake? Are the carbs in chips as bad as the carbs in a slice of wholemeal bread? Yes, all carbs are 'equal', but the issue is that it is easier to overeat carbs in cake, chips and bread than it is to overeat carbs in broccoli. Let's now take a quick look at how many carbs are in these various foods – a portion of chips has roughly 25g of carbs, a slice of cake 40g, two slices of bread have 40g and a large helping of broccoli has 6g (in 100g portions of each food). So you'd have to eat at least five large portions of broccoli to get the same amount of carbs as in one portion of chips, cake or bread. Much easier (and tastier) to eat a piece of cake. Now I know why biscuits are more addictive than broccoli. And why it is easier to eat three slices of cake than it is to eat three steaks in one go.

But besides the issue of the amount of carbs, there is yet another problem with cake and bread – if they contain wheat, then they will also contain some gluten and lectins (which are types of protein). And gluten and lectins come with a whole host of other problems.

GLUTEN: A DIVISIVE ISSUE

Many think that gluten is only a problem for those who suffer from coeliac disease. The thing is, consuming large amounts of gluten can be a problem for everyone. This is because gluten activates zonulin (a hormone), which in turn opens up the spaces between the gut lining cells. It is normal for these spaces to open a little in order to let nutrients into our bodies, but if there is too much zonulin, the spaces between gut cells open up too much. If you think of the gut as a pipeline, too much zonulin puts very big holes into it and allows substances that shouldn't be entering our bodies to come in. We will cover the problems that these blackballed substances cause in chapter 13.

LECTINS – THE EVIL RELATIVE OF GLUTEN

Lectins are found in almost all foods. But plant foods contain much higher amounts of lectins than animal foods. (Plant lectins are thought to play a role in the plant's defence mechanisms against being eaten.) Not all lectins are bad and some are more harmful than others. Most lectins are immune to stomach acid and digestive enzymes, and so are difficult to destroy if eaten raw. Like gluten, large amounts of lectins are also problematic. Harmful lectins inflict their damage via a variety of mechanisms – they cause a disruption of our appetite control mechanisms plus they act as 'anti-nutrients' by making it difficult for the body to make use of other nutrients. There is a subclass of lectin, which goes by the name of 'wheatgerm agglutinin', and is one of the most toxic of all the lectins (and it is only found in wheat). So it doesn't matter whether you are a coeliac or not, wheat, which is high in carbohydrates, gluten and lectins, causes problems for everyone. It is the level of wheat consumption at which problems start to occur that varies from person to person. In coeliacs it is minuscule amounts, while in others it is higher amounts. These are some of the reasons why cereals are the worst of all the plant foods.

TREATING DIABETES, OBESITY AND CHOLESTEROL WITH BETTER DIET

Having shown the evidence of how high-carbohydrate diets lead to diabetes and weight gain, we now summarize how the Sapiens diet can address those conditions.

Can diabetes be treated with diet? Yes, it can. Just as a diet with too much carbohydrate or sugar causes diabetes, alternative diets can remedy the condition. Many people have been able to come off all their diabetes medication by changing diet. Weight-loss regimes, low-carbohydrate diets and ketogenic diets have all been shown to reverse diabetes and to lower abnormally high insulin levels. The bad news, though, is that once full-blown diabetes has developed, it cannot be cured with diet, only controlled. While those who are prone to diabetes can reverse their condition by changing what they eat, if they return to eating in their old ways (excess carbohydrates or calories), they will eventually experience a recurrence of diabetes. Just as medication is effective only while one is taking it, a low-carb diet can only control full-blown diabetes as long as you stick to the diet.

With respect to obesity, the Sapiens is ideal for losing weight, and then keeping it off. On the Sapiens diet our innate appetite control mechanisms work as they should and there is much less temptation to overeat. Plus, being low or free of carbohydrates, the Sapiens diet doesn't stimulate excess insulin, and glucagon production isn't suppressed. This allows the body to burn the fat in one's diet – or once that's used up, to burn body fat as its energy source. So, fat gets burned, not stored.

The animal fats consumed as part of the Sapiens diet provide us with energy and nutrients. They are the fats we are biologically adapted to; indeed, they drove our evolution. It really isn't plausible that they've somehow become toxic to us in the last few decades. The reasons why animal fats are healthier than vegetable oils will be covered in chapter 10.

CONCLUSION

Some 80% of the world's population now have abnormally high insulin levels and close to 10% are diabetic. For the world to become healthier, we need to reverse these trends by reducing our intake of carbohydrates, switching to something more akin to the Sapiens diet. Carbohydrates are not inherently bad, but people who are prone to becoming overweight or diabetic have a defect in handling them. The simplest way of correcting abnormally high insulin and glucose levels is to reduce the amount of carbohydrates in one's diet, starting with the carbohydrates from grains because these foods also cause a leaky gut and contain anti-nutrients.

The evidence that overconsumption of carbohydrates and sugar is linked to diabetes, obesity and heart disease is overwhelming. Yet the skewed logic of current nutrition thinking, embodied in the UK Eatwell guidelines and the US food pyramid, leads us down the path to ill-health by promoting high-carbohydrate consumption to provide calories through whole grains, fruit and vegetables, rather than nutrient consumption from the animal foods we have become adapted to consume over millions of years of our evolution.

We evolved with the Sapiens diet. With carbohydrates we risk eating ourselves to extinction.

But hold on, some readers might be thinking, aren't we forgetting something? There's more to grains than calories and carbs – what about fibre? So let's talk about fibre next.

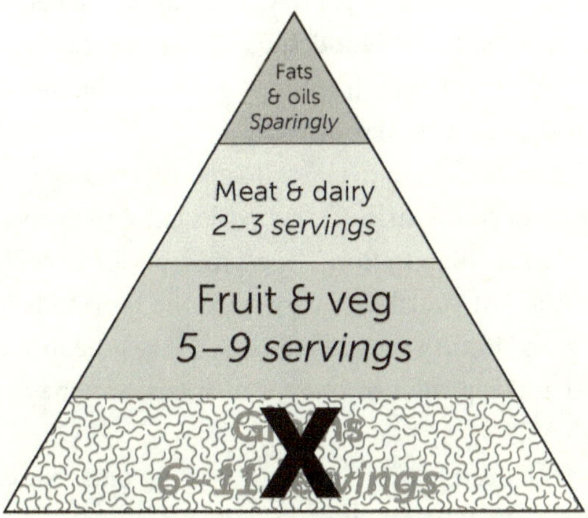

CHAPTER 6

FIBRE – A NATURAL JUNK FOOD

Fibre has no nutritional value

'The role of dietary fiber in constipation is analogous to cars in traffic congestion. The only way to alleviate slow traffic would be to decrease the number of cars and to evacuate the remaining cars quickly.'

Dr Ho Kok Sun (president of the Society of Colorectal Surgeons (2014–2016), Singapore)

FIBRE HAS GAINED QUITE A REPUTATION IN recent years. An internet search on fibre will provide a staggering number of articles on how to reach the recommended intake of 30g per day. The claimed benefits of consuming fibre include weight loss, lower blood sugar, lower cholesterol, constipation relief, reduced bowel cancer risk, improved gut microbiome, and protection against diabetes and cardiovascular disease.

And it doesn't require a Google search to get this message. Just take a stroll along any supermarket cereal aisle. Cereal boxes extol the supposed virtues of fibre, with labels such as 'high in fibre', 'high in superior fibre', 'fuels a healthy gut' and 'helps you feel at your best every day'. The food industry has bought into the fibre story, and in turn they sell it to us. 'Fibre is healthy and necessary' has been reported and repeated so often that it has become accepted as fact. Indoctrination works.

It would seem the need for fibre in our diet is beyond challenge.

We beg to differ. We'll begin by explaining what fibre is and then we will counter the two main claims for fibre: that it protects against bowel cancer, and that it relieves constipation.

FIBRE (AND ALL THAT GAS)

What we know for sure about fibre is that it is only found in plant foods – vegetables, grains, legumes, fruits, nuts and seeds. We also know that fibre cannot be digested by the human body. As mentioned in 'Nutrition 101' (in chapter 4) there are actually two main types of fibre: soluble fibre and insoluble fibre. Insoluble fibre is supposed to be beneficial because it *speeds things up* in the gut. And soluble fibre is supposed to be beneficial because it *slows things down* in the gut. Hmm …

The other supposed benefits of soluble fibre are that it makes us feel full for longer and also slows down the absorption of sugar, which in turn reduces sugar spikes in the blood. But wait, this list sounds familiar. Eating fat also makes us feel full, slows down the absorption of sugar and reduces sugar spikes – and it does so more than any other food. Why is it that all the kudos goes to fibre? Fibre is undigestible and provides no nourishment – it is totally unnecessary for the maintenance of life or for growth – why eat fibre? And if we do, what exactly happens to it as it passes through the gut?

Because we lack the enzymes needed to digest it, fibre reaches the colon (the large intestine) unchanged. Here, fibre gets fermented by the bacteria naturally living there. One of the side effects of this fermentation is the production of something known as 'short-chain fatty acids' (more on this later). Another, rather inconvenient side effect of fermentation is the production of gas. This only serves to cause bloating, which is unpleasant for the individual, and flatulence, which is unpleasant for others in the vicinity. One of the many reasons to beware large bowls of lentils. Another reason, which not many think about, is also the extra strain on the sewage system. Yet more reasons to beware fibre will be covered in the next chapter.

FIBRE AND CANCER OF THE BOWEL

Bowel cancer matters; worldwide, it's the third most common cancer – and expected to increase by 60% by 2030. One of the commonest misconceptions about fibre is that it reduces the risk of bowel cancer. The media, and even organizations such as Cancer Research UK, claim that dietary fibre can reduce the risk of bowel cancer. But where and how did the whole fibre story start?

Some fifty years ago, a British missionary surgeon, Dr Denis Burkitt, studied the differences in patterns of diseases between poverty-stricken people in Africa with diseases of affluence that are found in the West. What Dr Burkitt found was that people who followed a traditional African diet had lower rates of bowel cancer.

He also noticed that their diets were generally much higher in fibre. Based on these observations, Dr Burkitt suggested that a high-fibre intake reduces the risk of bowel cancer. He even went on to write an international bestseller, *Don't Forget Fibre in your Diet.* Dr Burkitt's theory was similar to the hypotheses that are generated in observational studies (which were discussed in chapter 3). These hypotheses need to be tested in controlled trials to see if they stand up. And often they don't.

Out of curiosity, we did a Google search using 'bowel cancer fiber' (deliberately using the American spelling) and came up with 12 million results. A similar search on Google Scholar (the part of Google that gives you only scholarly literature) came up with 147,000 results. It was difficult to know where to start, so we have picked some of the largest and most well-known studies and summarized them below, while noting that most of them are low-quality, epidemiological studies that cannot prove anything definitively. What really stands out is the bias shown in the researchers' comments. And even when their results suggest that fibre is not protective against bowel cancer, they often dodge and weave to account for their findings in ways that defend dietary fibre.

- The largest ever study on fibre and bowel cancer found no benefit from eating fruit, vegetables, cereals or other high-fibre foods. This was the Nurses' Health Study, which followed 88,757 women over sixteen years. Unexpectedly, they found that greater consumption of vegetable fibre was associated with a 35% increased risk of bowel cancer. Dr Charles Fuchs, a medical oncologist and lead author of the study, is quoted as saying, 'The effect was small but statistically significant.' (Statistically significant means that the finding is extremely unlikely to be a chance finding.) But despite being 'statistically significant', Dr Fuchs then goes on to dismiss the result by saying that it was probably a result of chance! He ends with, 'We really can't substantiate a protective effect of fiber on the rates and risk of colon cancer. That is not to say that fiber is not a good thing. It is still helpful, but not for colon cancer.'
- A study on 41,837 women conducted by the Johns Hopkins institute in 1994 also noticed an increased risk of bowel cancer with greater consumption of certain vegetables – for instance, brassicas (like broccoli and cauliflower) and also potatoes. A quote from the study: 'Surprisingly, intakes of cauliflower and potatoes appeared to be positively associated with colon cancer risk. It is notable that neither energy, fat, nor protein intakes were positively associated with risk in the present study.'
- Moving on to studies on men, the Health Professionals Study involved looking at the diets of 47,000 men over ten years. This concluded that there was no association between fibre intake and the risk of bowel cancer.

- Another large study, on 62,000 men and 70,000 women this time, found that higher intakes of plant foods or fibre (from whole grains, fruits and vegetables) were not associated with a lower risk of bowel cancer.

WHAT DO OTHER EXPERTS SAY?

According to a research group at the Imperial Cancer Research Centre, fibre may *increase* the likelihood of bowel cancer. The mechanism they suggest is that 'fermentable fibre substrates can stimulate proliferation in the colon. The overall significance of this is unclear, but in view of the general lack of benefit in experimental human studies and the detrimental effects seen in animal models of carcinogenesis, caution in the use of nutritional fibre supplements is advisable'. What does this mean? It means that the researchers found that fibre induces cell division (or proliferation, as they call it), which in turn may be a precursor to cancer. This increased 'cell division' action of fibre is clearly borne out in animal studies.

It isn't only UK scientists who are unsure about the benefits of fibre. An American Gastroenterological Association review on bowel cancer and fibre concluded: 'given a lack of complete scientific evidence, it's difficult to advise patients with absolute confidence'. Despite this, they go on to say: 'Nevertheless, the guidelines in this review represent reasonable conclusions based on currently available data. Therefore, it is reasonable to recommend total fiber intake of at least 30–35 g/day.' The panel seem to have found it impossible to consider that the default advice to eat plenty of fibre might be wrong. The bottom line is that the evidence does not support claims that lack of fibre in the diet causes bowel cancer, either in humans or in animals.

Because the 'fibre is beneficial for bowel cancer' story is so entrenched, we needed to reassure ourselves further, and we looked into controlled studies on fibre and adenomas (precursors of bowel cancer). Surely these would prove that high-fibre diets are good at preventing these pre-cancerous bowel tumours?

- A study on more than 2,000 people showed absolutely no preventative benefit of fibre on the risk of adenoma recurrence. This randomized controlled trial was reported in the *New England Journal of Medicine* (one of the most prestigious medical journals in the world).
- More worryingly, taking ispaghula husk (commonly known as 'Fybogel' in the UK) may increase the risk of adenomas. Fybogel is one of the most commonly prescribed laxatives in the world, and it's concerning that no warning is given of this possible adverse effect in the *British National Formulary* (the standard formulary used by doctors in the UK for prescribing medicines).

- But it isn't only ispaghula husk that is under the spotlight; other types of fibre such as pectin, guar gum, carrageenan and alfalfa have also been associated with an increased risk of contracting bowel cancer, or else they have had no effect one way or another.

In case you aren't convinced, here are two more reviews of randomized control trials (the gold standard of trials).

- A Cochrane review (the pinnacle of reviews of studies) of controlled trials found that 'increasing fibre in a Western diet for two to eight years did not lower the risk of bowel cancer. Paradoxically, after four years participants receiving dietary fibre had higher rates of bowel cancer compared with the control group'. Another quote from this review: 'There is currently no evidence from Randomised Control Trials to suggest that increased dietary fibre intake will reduce the incidence or recurrence of adenomatous polyps within a two-to-four-year period.' ('Adenomatous polyps' are precursors of bowel cancer.)
- And from another review: 'On the basis of current data, there is little evidence to support the use of dietary fiber supplements to reduce the risk of colorectal neoplasia (bowel cancer).'

Overall, the weight of evidence suggests that *fibre does not offer protection against bowel cancer.*

So, what does increase the risk of bowel cancer? If one had to pinpoint more likely dietary associations with bowel cancer, then it's alcohol consumption and high blood-glucose levels that show the strongest links. Yet, alcohol doesn't come with dire warnings despite its very strong association with cancer of the bowel, breast, liver, mouth, throat, oesophagus and stomach. Nor do carb-loaded foods such as wholegrain bread, rice, granola and sugary drinks come with warnings, despite their clear role in pushing blood glucose levels up.

FIBRE AND CONSTIPATION

Now we come to another common myth about fibre. Constipation.

It is worth stating the obvious: defaecation is a side effect of eating; it is not the purpose of eating. Regardless, the obsession that some people have with their toilet habits is as old as the hills. Apparently, between 8% and 33% of people in the world are constipated. Since constipation appears to be rather common, we will now debunk the myth that a lack of fibre in the diet causes constipation.

First, let's start with a definition of constipation. A general definition goes along the lines of: 'a condition in which there is difficulty in emptying the bowels, usually associated with hardened faeces'. And according to the 'Rome criteria' (an international classification system), constipation can be diagnosed when two or more of the following occur:

- Straining at least 25% of the time
- Hard stools at least 25% of the time
- A feeling of incomplete evacuation at least 25% of the time*
- A feeling of anal blockage at least 25% of the time*
- Manual manoeuvres for rectal emptying at least 25% of the time*
- Two stools or less per week

* These symptoms indicate an impairment of defaecation.

Probably a bit too much information for some, but it is important to know exactly what constipation is if we are going to correct misassumptions. It is popularly thought that one should open one's bowels every day, and if this isn't the case, then most people think they are constipated. But it's only if someone goes twice or less per week that they can be considered constipated, and then only if another symptom is present as well. What many aren't aware of is that there is, actually, no minimum stool frequency required for good health.

The media, dieticians and doctors repeatedly tell us that fibre is a requirement for regular defaecation, that fibre is a necessary component of a healthy diet, and is needed for a healthy microbiome (we will look at microbiome misconceptions shortly). But are these beliefs correct? While medical textbooks say that fibre improves the passage of food through the gut, there is actually no proof of this in practice nor in experiments.

A team of medical researchers decided to investigate fibre and constipation in a very well-conducted study. What they found was that stopping all fibre intake also stopped constipation. And it wasn't just some of the patients who stopped being constipated – it was every single one. These researchers aptly used a traffic jam as an analogy to the role of dietary fibre in constipation. An obvious way to alleviate traffic congestion is to decrease the number of cars joining the jam. Or to let cars get out. Letting more cars join the traffic jam will only add to the problem. Similarly, if a bowel is already packed with faeces, consuming extra fibre will only make emptying the bowel more difficult. Basic physics and several excellent studies affirm this finding. (Humans cannot digest fibre, so all the fibre we eat has to be excreted.) The researchers of the aforementioned study titled their paper, 'Stopping or reducing dietary fiber intake reduces constipation and its associated symptoms'. They concluded that we should re-examine popular beliefs about the benefits of fibre. However, the idea that increasing the volume of stools will ease the process of passing a stool is still taken as a given.

Most people who seek medical help for constipation are usually already consuming prodigious amounts of fibre. This explains why surgeons find the colons of severely constipated people are usually packed with undigested vegetable fibre. These findings, plus a bit of common sense, suggest that adding further bulk to the hard and large stools of the chronically constipated would not be a wise thing to do. Clearly, common sense is not that common – despite all this evidence, it remains standard medical (and layperson) advice to treat constipation by increasing fibre intake by consuming yet more fruit, vegetables and cereals. In fact, the best advice is to *decrease* consumption of all fibre containing foods. And remember that you don't need to go every day!

IS FIBRE NEEDED? EVOLUTION SAYS 'NO'

As always, when considering the effects of a particular food on human health, it pays to look at things under an evolutionary lens together with the underlying biological mechanisms.

The giveaway is our guts. Nothing happens overnight in evolution; the reshaping of the gut commenced 6 million years ago, which is when the human lineage split from the chimpanzee line. As mentioned in chapter 2, our brains and guts co-evolved; eating more and more animal foods led to our guts becoming structurally closer to the guts of carnivores. Now, compared to chimpanzees (and all other primates), humans have a smaller gut with a relatively long small intestine and a short colon (large intestine). The surplus energy created by a reduced-size gut generated the extra energy needed for our brains to grow.

The average length of the human small intestine is twenty-three feet (it is small only in diameter), nearly five times longer than the human colon. By contrast, in chimpanzees and other apes the small intestine and colon are about equal in length. So, what is the significance of this? This configuration of the human gut is more like that of a carnivore than that of a herbivore. Carnivores need a long small intestine because it is the organ responsible for digesting and absorbing nutrients from animal foods. For example, 90% of all protein eaten is digested in the small intestine. The human small intestine has lengthened during evolution because we evolved to eat meat, whereas apes have evolved a relatively short small intestine because they didn't. But the small intestine cannot break down fibre.

The apes have evolved to rely on a very fibrous diet of plant foods. And since such foods have a low nutrient density, the ape must chomp through vast quantities every day. All this fibre can only release its nutrients after it has been broken down by bacterial fermentation. This happens in the colon, and hence the apes have evolved a large colon, whereas our low-fibre, meat-heavy diet enabled evolution to shrink the human colon. Thus, the human gut is now incapable of breaking down the stupendous quantities of fibrous foods that other apes are obliged to routinely consume.

IS FIBRE NEEDED? BIOLOGY ALSO SAYS 'NO'

Many scientists are of the opinion that fermenting fibre is one of the important jobs of the colon (and remember that 'fermentation' is not the same as 'digestion'). This is because, besides producing gas, the fermentation of fibre also yields substances known as 'short-chain fatty acids'. And many of the supposed benefits of a high-fibre diet are down to these short-chain fatty acids (which are a type of fat). Let's now take a look at what goes on in the colon and what these short-chain fatty acids are all about.

After spending roughly two hours in the small intestine, only leftover unabsorbable food, such as fibre, is expelled into the colon. Very little, in terms of digestion and absorption of nutrients, goes on in the human colon; the main job of the colon is to reabsorb water. The colon, however, is good at doing its job and it doesn't need the help of fibre to do it. The moisture content of faeces is kept constant (between 70% and 75%, in case you were wondering) regardless of the amount of fibre consumed. It is a complete fallacy that fibre aids water retention in the colon and makes faeces easier to pass. It is also a fallacy that our colons need to be swept clean by fibre – the colon cells renew themselves automatically and completely every few days, regardless of what one consumes. You can cancel that colonic irrigation appointment …

Which brings us to short-chain fatty acids, another supposed benefit of fibre. These are, purportedly, 'anti-inflammatory, immunoregulatory, anti-obesity, anti-diabetes, anticancer, cardiovascular protective, and neuroprotective'. An impressive list. But as we will learn shortly in chapter 11, the human body makes ketones naturally when following a ketogenic diet like the Sapiens diet. And ketones are also 'anti-inflammatory, immunoregulatory, anti-obesity, anti-diabetes, anticancer, cardiovascular protective, and neuroprotective'. Furthermore, ketones are more potent than short-chain fatty acids.

COMPARING KETONES AND SHORT-CHAIN FATTY ACIDS

Fermentation of fibre produces many different short-chain fatty acids. Let's now compare the main short-chain fatty acids that are produced by fermenting fibre with the main ketones that are produced on a ketogenic diet.

- Main short-chain fatty acids: *butyrate* and *acetate*.
- Main ketones: beta hydroxy*butyrate* and aceto*acetate*.

Almost identical.

Yes, we know that a tiny difference between two molecules doesn't necessarily mean that they have similar effects, but in this case the small difference means that the short-chain fatty acids get converted easily into ketones. The cells of the colon convert at least 70% of short-chain fatty acids into ketones (for example, butyrate is made into hydroxybutyrate). Or you could say that many of the benefits of short-chain fatty acids are exerted when they are converted into ketones.

So why is it that when the media tell us that we should all be consuming more fibre, they don't also tell us that quite a few of the benefits of fibre are mediated by ketones? Nor do they broadcast the downsides of short-chain fatty acids, which is even longer than the list of their benefits. For example, short-chain fatty acids can induce brain neurochemistry impairment and promote fat storage in the liver (one of the worst places to store fat). Studies on mice may show benefits of short-chain fatty acids, but studies on humans show the complete opposite – I would go with the studies on humans.

Why suffer side effects of short-chain fatty acids when we can get all the benefits, and more, from ketones that our bodies make when we follow a ketogenic diet? All the supposed benefits of short-chain fatty acids derived from fibre also apply to ketones, but without any of the downsides. The bottom line is that if you are consuming a ketogenic diet, you will have a readymade supply of ketones and there is no need to rely on the conversion of short-chain fatty acids into ketones.

Plus, according to researchers, 'more high-quality studies are needed to verify the effects of short-chain fatty acids on humans'. A bit like the microbiome.

EVERYTHING YOU WANTED TO KNOW ABOUT THE MICROBIOME BUT WERE TOO AFRAID TO ASK

The current fad for feeding our gut microbiome is a good example of a distraction in dietary health; it's a wonder we've survived these last few millions of years and become the dominant species on the planet without the benefit of prebiotic supplements, probiotic supplements and dietary advice from newspaper supplements.

So, what exactly is the microbiome?

The microbiome consists of all the microorganisms (such as bacteria, viruses and fungi) that live in our gut. Certain types of gut microorganisms are believed to be beneficial and some to be harmful. However, most of the evidence regarding the gut microbiome is unreliable, since the majority of studies are not controlled studies (the best type of studies). And what about the controlled studies? As with fibre and short-chain fatty acids, controlled studies on the human microbiome are inconclusive. It is the observational studies (a poor type of evidence) that suggest that the gut microbiome may be of any relevance. But despite the lack of convincing evidence, the microbiome mantra has become all the rage.

Where do the claims made for fibre helping maintain a healthy gut microbiome come from? Because different types of fibre are processed by different species of microbes, we are urged to include a wide variety of fibre-containing foods to keep all these microbes happy. And it seems these bugs are particularly fond of fermented foods – which is why the chattering middle class like to keep up with their kefir, kombucha and kimchi.

So how exactly do we benefit from all these microbes zinging around in our gut? Surely there must be some concrete reasons for all the healthy microbiome hype? According to some, these microbes strengthen our gut wall and prevent a leaky gut. But to date, not a single study has shown that changing the microbiome by itself is responsible for reversing leaky gut. Instead, there is evidence to the opposite; consuming large amounts of fibre can *cause* a leaky gut. (This is partly due to the excessive amounts of short-chain fatty acids that are produced from the fermentation of fibre in the large intestine.)

Which brings us to one of the most important topics of this book – why a healthy and uncompromised gut wall matters. This topic of 'leaky gut' will be covered in detail in chapter 13. Suffice it to say for now that the evidence of the harms of a leaky gut is rock solid, unlike the evidence for the benefits of the microbiome. Gut-barrier integrity is not an outcome of 'maintaining the health of your microbiome' – there is no need to conflate the two. In fact, in susceptible people, certain plant foods prevent the proper functioning of the gut barrier; in such cases fibre-containing foods are likely to be harmful.

What *should* be conflated with improved gut health is a Sapiens-type diet. This has consistently been shown to be of benefit in serious gut diseases and *always* reverses a leaky gut. And when a leaky gut is reversed, there are clear improvements in health. Indeed, the repairing of the gut barrier is one of the central mechanisms that explains the improvements in health that occur on the Sapiens diet (the other main benefit is down to an improvement in insulin regulation). If you had the choice of friendly bacteria in your gut or an intact gut barrier, which would you take? On the Sapiens diet, your gut barrier will be intact and your body will also be producing ketones, so there is no need to worry about having to eat fibre or which bugs exactly have taken up residence in your gut.

Now we will move on to the links between fibre and other serious, but non-cancerous, bowel diseases and then take a brief look at the links between fibre consumption and body fat. We will end by looking at what happens to breastfed babies, as nature has dictated that they have absolutely no fibre in their diet.

FIBRE AND OTHER DISEASES OF THE BOWEL

Fibre consumption is frequently recommended for the prevention of some serious bowel diseases, yet during flare-ups of these very same diseases, one is cautioned *against* consuming fibre. The evidence that fibre is harmful during flare-ups in ulcerative colitis and Crohn's disease is incontrovertible. When considering the risk factors for these serious bowel diseases, an American paper says, 'During inflammatory bowel disease exacerbation periods, recommending a low fiber diet is unquestionable.' Surely this would be a fine time to reflect on whether their original premise, 'a high fiber diet may play a protective role in inflammatory bowel disease', could be wrong? It seems bizarre that fibre can help prevent a serious disease and also worsens it when the disease flares up.

FIBRE AND BODY FAT

Visceral fat is fat that is found around one's internal organs and it increases our disease risk. But what has fibre got to do with visceral fat? Well, one study looked at the correlation between amounts of dietary fibre consumed and levels of visceral fat. The researchers fully expected to find that the more fibre you eat the less visceral fat you have. They measured the visceral fat using MRI scans, which are very accurate, so there was no uncertainty. They found that the highest calorie diets were associated with the highest levels of both types of fat – visceral fat and subcutaneous fat. The more you eat, the fatter you get. No surprises there.

However, unexpectedly, this study found that those with the highest fibre and highest carbohydrate intakes were the ones with the highest levels of visceral fat. The authors of this paper say that it's hard to give a good explanation for this finding. Once again, they, like others we've already encountered, dare not challenge the dogma that fibre is beneficial. To the authors of this book the study adds weight to the evidence against the current dietary recommendations. Even those who do as they are told and consume a high-fibre and high-carbohydrate diet are at an increased risk of visceral fat.

FIBRE AND BABIES

Let's conclude by looking at breastfed babies, who are a perfect example of a no-fibre diet. Given the fact that their diet has the blessing of nature, we can see whether what we have shown in adults can be confirmed by looking at constipation, the microbiome and short-chain fatty acids in breastfed babies.

As far as constipation goes, what we find is that breast milk contains absolutely no fibre and babies who are exclusively breastfed *never* get constipated. Babies have absolutely no need for fibre to make their bowels move easily. In fact, giving fibre to babies can be downright dangerous and sometimes cause a life-threatening disorder.

If we now move on to the microbiome, we find that babies are born with no bacteria in their guts; that is, they have *no* gut microbiome at birth. The process of gut bacterial growth takes time in babies. But the process of gut bacterial growth takes *nine* times as long in breastfed babies as it does in formula-fed babies. It's odd that if there were advantages to having more gut bacteria that evolution hasn't made sure that breastfed babies more rapidly acquire these bacteria.

And if we compare the levels of short-chain fatty acids in breastfed and bottle-fed babies, we find that the former have half the levels that are found in formula-fed babies. In fact, extremely high levels of short-chain fatty acids in babies are associated with necrotizing enterocolitis, a fatal disease caused by an inflamed intestine. Given the fact that breastfeeding is healthier for babies than formula feeding (providing that the mother is well nourished), one could infer that the lower the level of short-chain fatty acids, the better it is for babies. (In necrotizing enterocolitis damage to the gut cells allows bacteria to invade, giving rise to a mortality rate of 50%.)

While we appreciate that the diet of breastfed babies is not representative of an adult's diet, the point is that a lack of fibre does not cause issues with the microbiome of babies, nor does a lack of fibre cause constipation. As far as short-chain fatty acids go, breastfed babies appear to function better with lower levels of these substances, probably because they are efficient at making their own ketones. The bottom line is that babies don't need fibre and large amounts of fibre are actually dangerous for them.

SUMMARY

There are many harms to eating too much fibre – bloating, flatulence, discomfort, constipation and in extreme cases intestinal blockage. One of the benefits of eating grains is supposed to be owing to the fact that they are a good source of fibre. But our bodies don't need fibre, neither from grains nor from other plant foods. Evolution says we don't need fibre. Our biology says we don't need fibre. Babies don't need it, and neither do adults. As a species we are adapted to a *low-fibre*, animal-based diet.

Which means the change we made to the food pyramid in the previous chapter – reducing carbs – is further corroborated by discounting fibre. Coming up next for demolition: fruit and vegetables.

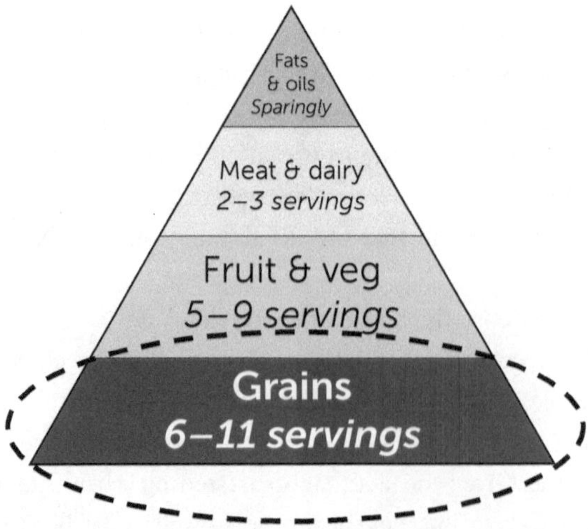

CHAPTER 7

VITAMIN C AND THE FRUITLESS DIET

Animal foods provide all the vitamins we need, including vitamin C

'Be a free thinker and don't accept everything you hear as truth. Be critical and evaluate what you believe in.'

Aristotle

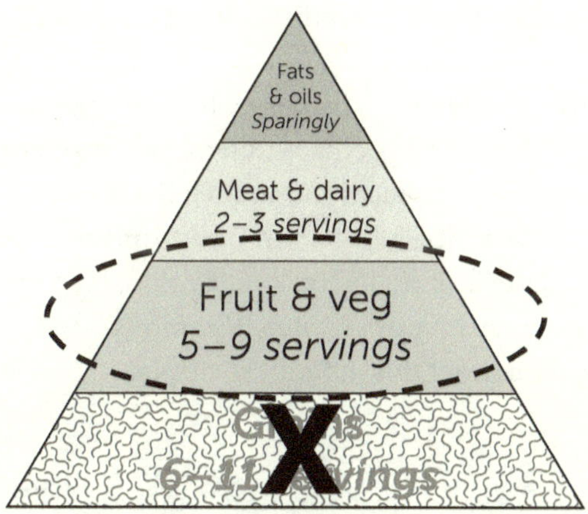

IF SOMEONE HAD SAID TO ME A few years ago that I would one day be cutting out fruit and vegetables from my diet, I would have been horrified. But that day did come. It has now been nine years since I gave up eating plant foods. And, yes, I am still here to tell the tale. Without scurvy or constipation. In fact, forgoing cereals, fruit, vegetables and all other plant foods has improved my health no end.

Both the food pyramid and the UK's Eatwell plate recommend the intake of substantial quantities of fruit and vegetables, so much so that on the Eatwell plate it's the largest section. In the UK fruit and vegetables are often promoted with the slogan 'five-a-day'. But that's not quite accurate – the advice isn't 'five-a-day', it's *at least five* portions of fruit and veg a day. Five is the lower limit. But it isn't just the UK – most developed countries (such as the USA, France and Germany) also recommend at least five-a-day and they all base their recommendations on advice that originated from the World Health Organization.

As with fibre, there may be those readers incredulous that we might be about to recommend ditching an entire food group, and perhaps doubtful of our claim to be healthy having done so ourselves. 'What about vitamin C?' might be one question. 'Don't we depend on fruit and vegetables, especially citrus fruit, for that vitamin?' Which is why, in this chapter, as well as discussing the nutritional unreliability and drawbacks of fruit and vegetables in general, we'll first address vitamin C specifically.

VITAMIN C

Vitamin C. For many it's the vitamin they are most familiar with. It's the vitamin that comes to mind first, say, when thinking about the nutrients in our (at least) five-a-day of fruit and vegetables. It's the vitamin we often turn to when we feel under the weather, perhaps suffering a cold or flu, in the belief that it will shore up our body's defences and shorten the illness.

Also known as ascorbic acid, vitamin C performs several vital roles in the body. It protects our cells, it maintains healthy skin, blood vessels and bones, and it promotes the healing of wounds. Vitamin C is also an antioxidant, though we'll say more about this in the next chapter.

VITAMIN C – A BIT OF HISTORY AND BIOLOGY

Vitamin C is needed in the diet because our ancestors lost the ability to make it for themselves some 50–60 million years ago, when one of the genes involved in the synthesis of the vitamin from glucose became inactive. At around the same time we also lost the ability to break down uric acid, resulting in higher levels of uric acid in our bodies. Because uric acid is a more potent antioxidant than vitamin C, it is thought that uric acid took over where vitamin C left off.

A bit more biology. For vitamin C to get into a cell, it needs to cross cell membranes using 'transporters'. What isn't very well known is that vitamin C and glucose use the same transporter to get inside cells. And when glucose levels in the blood are high, glucose hogs the transporters. In such situations, vitamin C has difficulty using the transporters, and has difficulty getting into cells. This means that most of the vitamin C gets stuck outside the cells. Trouble is, inside the cells is where vitamin C is actually needed.

These are the reasons why higher than normal intakes of vitamin C are of no benefit to those with constantly high blood-glucose levels (such as those with diabetes). However, all these vitamin C problems are irrelevant in diets that are low in carbohydrates – vitamin C wins the competition with glucose and finds its way easily into cells.

A SHORT HISTORY OF SCURVY – THE SCOURGE OF SAILORS

Most know that scurvy is caused by a lack of vitamin C. What isn't common knowledge is that it takes roughly a month of absolutely no vitamin C before we start getting symptoms of scurvy. And it takes a full six months without vitamin C before severe consequences of scurvy start kicking in. Either way, it's not a route we would recommend – prolonged deficiency of vitamin C eventually causes death from infection or bleeding. Which is why scurvy used to be a big problem for sailors on long sea voyages.

During the age of sail, in the fifteenth to seventeenth centuries, more sailors lost their lives to scurvy than to war, shipwreck or disease all combined. Sailors' diets usually consisted of salted beef, beer and biscuits. On Magellan's three-year voyage, the crew numbered 250 when they set off and was whittled down to eighteen by the end. Like most people, I used to assume that one could only get vitamin C from fruit and vegetables and that the reason why the sailors of yesteryear succumbed to scurvy was because they had no access to fresh food on their long voyages. This idea was reinforced by the 'Limeys' story – the nickname given to the British sailors by the Americans. When it was first discovered that citrus fruits could ward off scurvy, the British navy started giving their sailors limes or lemon juice rations on the long voyages.

It therefore came as a complete shock to me when I learned that it is eminently possible to get vitamin C from animal foods. In fact, all offal is a good source of vitamin C. This explains why, well before the discovery of vitamin C, and before limes were taken on board ships, there were a lucky few on those long sea voyages who escaped scurvy. The officers on board were provided with limited amounts of egg, milk, fat and fresh meat from live animals on their ships. The lowly sailors, though, did not have access to any of the produce from the live animals on board. These rations meant that many of the upper echelon officers avoided the scourges of scurvy; the majority of the scurvy deaths were in the low-ranking sailors below decks. Animal foods must have provided enough vitamin C to prevent scurvy in the lucky high-ranking officers.

Similarly, during the American Civil War, soldiers' rations consisted of salted beef, corn and dried beans, which led to scurvy and eventually the death of many soldiers. Only those who were able to forage the countryside for fresh meat, fowl or eggs survived the ravages of scurvy.

Hunter-gatherer tribes have long known the value of offal. Indigenous Australians, the Inuit, the Masai and most other hunter-gatherers prize the offal of animals above all other foods. After a kill, it is the offal that is shared out first – these cultures make sure that everyone in their group gets a piece of offal when they sit down together for a meal, no matter how small the total amounts are. For example, the adrenal glands of moose, which are especially high in vitamin C, used to be divided between the entire Inuit tribe in the Yukon even though the glands are relatively small. The bottom line is that minute pieces of offal provide us with all the vitamin C that we need.

OTHER DISEASES AND VITAMIN C

Back to the present day and modern societies. Vitamin C certainly saves us from scurvy, but nowadays other grand claims are also made for it – like the suggestion that it's associated with a lower risk of cardiovascular disease and cancer. However, it is again important to remember that association does not mean causation. For every study that shows vitamin C to be beneficial, there is another one that shows no such thing. For example, a report that pooled together many individual studies to give a total sample of 350,000 women showed that eating fruit and vegetables, with all the vitamin C they contain, was not associated with a reduced risk of breast cancer.

THE FIVE-A-DAY FALLACY EXPOSED

We are advised to eat at least five portions of different-coloured fruit and vegetables a day. But why at least five? And why different colours? Here's what Eatwell says on the UK NHS website:

> Fruit and vegetables are a good source of vitamins and minerals, including folate, vitamin C and potassium.

Yes, they are a source of some vitamins and minerals. But are they a good source?

The following figures are comparisons of the vitamin and nutrient content of 100g of beef liver versus 100g of blueberries (an often-touted 'superfood'), and 100g of broccoli (two graphs because of the different scales). Hopefully, it's clear from this comparison which is the true superfood. Hint – it's not blueberries or broccoli.

Broccoli is indeed good for vitamin C and some other nutrients, but is completely rubbish for selenium, zinc, haem iron, vitamin A or vitamin B12 (there is absolutely no vitamin A, vitamin B12 or haem iron in broccoli or in blueberries). However, even though the amount of vitamin C in liver is lower, it is still adequate, and, as we shall shortly see, is much more than adequate on the Sapiens diet.

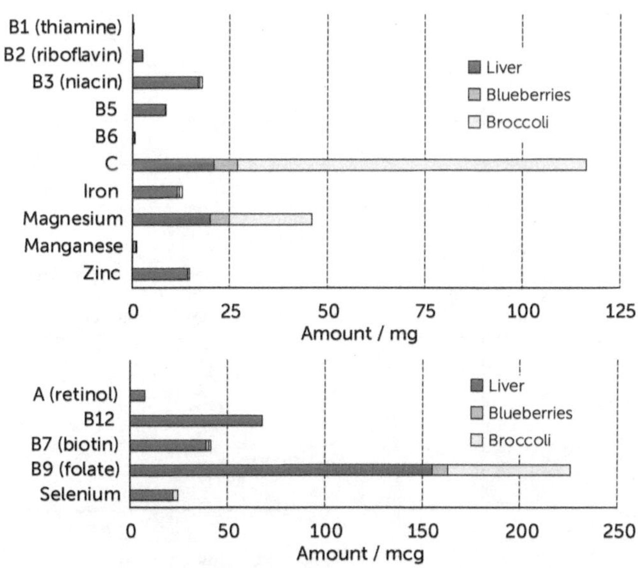

Nutrients in 100g of beef liver, blueberries and broccoli.

ANTI-VITAMINS AND BIOAVAILABILITY

The story gets even worse. The gross amounts of vitamins and other nutrients that fruit and vegetables contain – as would be normally indicated on the back of a packet – is not the same as the net amount absorbed by the body. Many of the nutrients in plant foods are not 'bioavailable', because they contain anti-nutrients that limit the uptake of vitamins and nutrients. Why do plants contain anti-nutrients?

Like all other living organisms, the aim of plants is to survive and multiply. Because plants can't physically fight or flee from their predators, they have evolved devious ways of defending themselves. Some of their physical defences include thorns and prickles, to make themselves less attractive to grazing animals. But the most powerful of their defences – anti-nutrients – are hidden. Some of these anti-nutrients include:

- Lectins: found in tomatoes, potatoes, goji berries, peppers and aubergines (eggplant), and in all legumes, such as lentils, black beans, kidney beans, soybeans, peanuts and chickpeas, and, last, but not least, in wheat (wholegrain wheat has the highest content of lectins).
- Oxalates: in green leafy vegetables, almonds, potatoes, tea, rhubarb and in grains.
- Phytates: in all grains, with much higher amounts in whole grains; in legumes; in nuts and seeds such walnuts, pine nuts, almonds and sesame seeds; and in tubers, for example potatoes, turnips, beets and carrots.

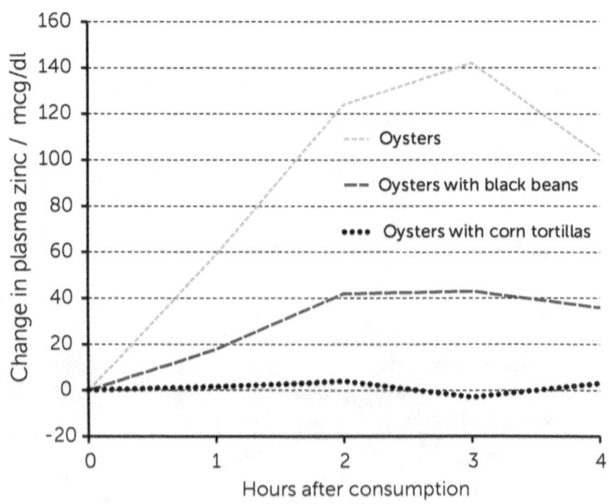

Bioavailability of zinc in oysters. Zinc readily absorbed when oysters eaten by themselves, but absorption plummets when eaten with beans or corn.

Bioavailability is a word rarely heard in the media. Instead, they are more likely to extol 'hearty and nutritious' vegetable casseroles and tray bakes. Hearty? That's subjective. Nutritious? Not as much as you'd think.

Back to the pyramid and the plate. They urge us to consume *at least* five-a-day of fruit and vegetables – which could mean seven or even ten portions, as some advocate. Why *at least* five if, as they also say, 'Fruit and vegetables are a good source of vitamins and minerals'? Five-a-day – sorry, *at least* five-a-day – is about *quantity* rather than *quality*. In fact, fruit and vegetables are relatively poor at providing vitamins.

Delving deeper into the guidelines, Eatwell acknowledges that fruit and vegetables are inadequate for vegans:

> For a healthy vegan diet have fortified foods or supplements containing nutrients that are more difficult to get through a vegan diet, including vitamin D, vitamin B12, iodine, selenium, calcium and iron.

Presumably those following a vegan diet are choosing their fruit and vegetables from the same range as everyone else, and are not just desperately unlucky in what they pick. Which means this admission is true for all of us – you need to be looking elsewhere for the missing vitamins and minerals. And where should we be looking? To meat and offal. But liver – one of the most nutritionally complete foods around – is not mentioned anywhere in the guidelines.

CARROTS: WHAT'S UP, DOC?

Eatwell could have included vitamin A in their list of missing vitamins, since no plant food contains any. But that fact doesn't stop Eatwell from endorsing supermarket carrots in the UK as 'rich in vitamin A' (see image on p. 106). Presumably they are thinking that the beta-carotene contained in vegetables such as carrots (or spinach, tomatoes, broccoli, peppers – in general, the more intense the colour of the vegetable, the more beta-carotene it has) is close enough, as it is converted into vitamin A by the body. Except there are two big problems.

Firstly, bioavailability again. Beta-carotene is not easily absorbed by the gut, especially in the absence of fat (and remember, Eatwell is a low-fat diet). One study estimated that only 11% of the beta-carotene in carrots ends up as vitamin A in our bodies if the carrots are eaten raw or boiled. But the bioavailability shot up to 70% if the carrots were stir-fried or roasted in fat. This is because beta-carotene is fat-soluble and eating it with some fat makes it easier

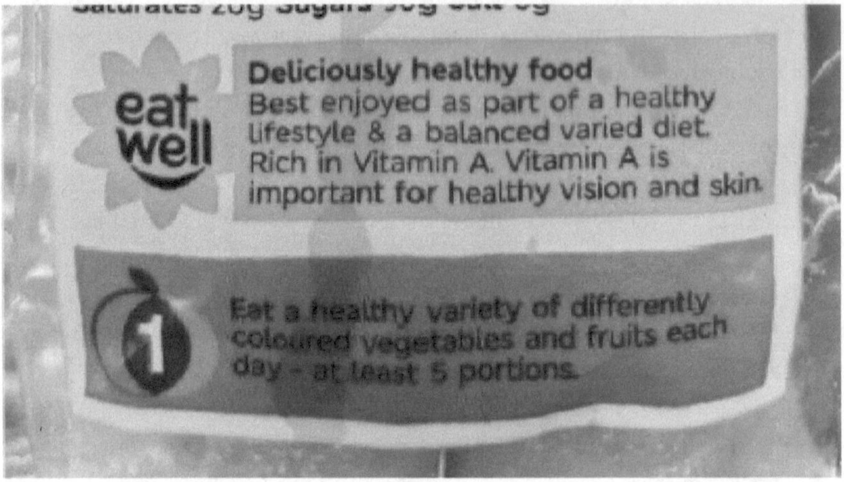

Dear Eatwell: there is no vitamin A in a carrot, or indeed in any vegetable, brightly coloured or otherwise.

to absorb. Actual vitamin A (also known as retinol) wins in the bioavailability stakes, typically 75–100% of vitamin A in animal foods is absorbed.

Secondly, around half of us do not have the right genetics to efficiently carry out the metabolism of beta-carotene into vitamin A. It's likely that many, many people, thinking they are eating healthily by piling on the grated carrot into their plastic container at the salad bar, are not getting the vitamin A they imagine. Because of genetics and the poor absorption and conversion of beta-carotene into vitamin A, if you are relying on beta-carotene for all your vitamin A, you need twelve times as much beta-carotene as you would ready-made vitamin A.

THE COSTS OF LIVING

It's not even the case that fruit and veg offer value for money. In September 2023 we did an online shop for fruit and vegetables. We assumed an adult seeking to follow the at-least-five-a-day guidelines might reasonably decide to aim for seven portions a day; that's nearly fifty portions of fruit and vegetables a week. For a family of two adults and two medium-sized children, that's 150 portions a week (a small apple is a typical portion).

One hundred and fifty portions of fruit and veg (total weight 12kg). Incomplete nutrition for £30.

Three trays of liver (total weight 1kg). More complete nutrition, just £8.

Let's spell that out for emphasis. One hundred and fifty portions for a family of four. One hundred and fifty portions of fruit and vegetables that Eatwell describes as a 'good source' of nutrition. How do Eatwell imagine we survived the last 300,000 years of our existence if we required such prodigious quantities of plants? Especially as plant foods contained much higher levels of anti-nutrients before we selectively bred them. We were hardly hanging about for supermarkets and global supply chains to emerge. But as we've said, five-a-day is about quantity, not quality.

Our order of 150 portions of fresh and frozen vegetables and fresh and canned fruit came to a total of £30 (total weight 12kgs). We then deleted the shopping basket and instead picked the alternative to 150 portions of fruit and vegetables – three trays of beef liver (total weight 1kg). That cost £8.

Thirty pounds for an inadequate source of vitamins, or £8 for a more complete source of vitamins? Which would you choose? And if you do your shopping in a physical supermarket, it's not just cheaper but also a lot quicker at the checkout.

Back online, we repeated the imaginary shop but substituting organically grown products where available. The price for fruit and vegetables shot up from £30 to almost £77. Whereas 1kg of pasture-fed liver can be bought for £10. More nutrition, sustainably farmed, for £67 less.

We also entered the shopping lists into a nutrition calculator, to compare the differences in nutrients. Here are the headlines:

- The fruit and veg on our list provided 6,500kcal. The liver 1,500kcal.
- The plants provided just over 1,000g of carbohydrates, the liver 0g.
- Of those carbohydrates, for the fruit and vegetables, 700g were sugars.
- The two had the same amounts of vitamins B9 (folic acid) and C.
- There was no vitamin A (remember, beta-carotene is *not* vitamin A), D or B12 in the fruit and veg. But plenty of all those in the liver.

The fruit and veg had 50g of iron, but this is not the easily absorbed form of iron, so a fraction of that amount would be absorbed by our bodies. Whereas 70g of haem iron, which is easily absorbed, is provided by liver.

Although the fruit and vegetables are poor in vitamins, they do manage to supply the best part of a large bag's worth of sugar for the family for the week. If I were told to eat a large bag of sugar every week, I would rather it were in the form of chocolate ... Besides the extra bag of sugar, all those portions of fruit and vegetables will add to unnecessary fibre in the gut.

This is a quote from a Public Health England webpage (from 2016):

> Only one in five of us meet the 5 A Day recommendation for fruit and vegetables, which is deeply concerning. PHE has redesigned the 5 A Day logo and made it free and easier to use on packaging. This will help consumers make healthier choices.

A logo. Phew, well that's all right then. Struggling to eat, or afford, five-a-day? Here's a logo. They say that they are deeply concerned, yet make no mention of considering a couple of portions of liver a week. Try searching the NHS website. Enter 'liver' and five pages of diseases of the liver are returned. But nothing about the benefits of liver as a food.

THE SAPIENS DIET AND VITAMIN C

One of the aims of this chapter is to show that it is possible to be healthy and obtain adequate vitamins, including vitamin C, on a diet that contains no plant foods. However, this is only the case if the Sapiens diet is followed; that is, the diet should consist mostly of animal fat, meat and offal. For those who have no health issues, one or two portions of fruit and vegetables can be added (with a maximum of 30g of carbohydrates/day). For those with serious health issues, the strictest form of the Sapiens diet is the best course of action; the strict version consists of animal fat, meat and offal, and nothing else.

It's the case that most current dietary guidelines recommend plant foods as the source of vitamin C. It's not just fruits like lemons and limes that contain vitamin C, so too do vegetables, particularly the colourful ones, such as peppers, tomatoes, broccoli and Brussels sprouts. A medium-sized orange contains around 50mg of vitamin C, and a 100g serving of broccoli around 90mg.

But offal, especially liver, is a good source of vitamin C. A 100g portion of liver has just over 25mg of vitamin C, about the same amount as in 100g of spinach. However, the vitamin C level in spinach is reduced by 70% when it's boiled, whereas the vitamin C in offal isn't degraded very much at all by cooking. For example, 100g of uncooked liver contains 25.3mg of vitamin C; if cooked, it delivers 23.6mg: 70% reduction vs 7% reduction. And because there aren't any other substances inhibiting its absorption, this all adds up to the fact that liver is superior to plants at providing vitamin C. And as with other fruits and vegetables, vitamin C is absorbed more easily and efficiently from animal foods than it is from plant foods; it is more bioavailable.

CONCLUSION

Vitamin C is an essential requirement in our diets, and a lack of it results in scurvy. The Sapiens diet provides more than enough vitamin C. But it goes further than vitamin C. Rather than ask if an animal-based diet can provide enough vitamin C, a better question would be to ask whether, apart from vitamin C, we can depend on fruit and vegetables to supply anything else? Beta-carotene from vegetables is an unreliable source of vitamin A, for example. And there's no vitamin B12 or vitamin D. There's no bioavailable iron. And anti-nutrients found in plants further limit how much vitamins and minerals we absorb. In contrast liver supplies a bioavailable range of the nutrients the human body and brain needs.

On the Sapiens diet, instead of five-a-day we would suggest one-or-two-a-day for those who have no medical problems and for those actually enjoy eating fruit and vegetables. However, we would recommend zero-a-day for anyone with any serious medical condition (more details on this in chapter 14).

We have UK dietary advice and a national health institution that fail to mention one of the most nourishing foods available. This is scandalous. Liver is a forgotten superfood. It's time to put it back on the shopping list. A piece of liver a day will almost certainly keep the doctor away.

We haven't entirely finished with fruit and vegetables yet; we also need to dispel another part of the undeserved aura that surrounds them – antioxidants.

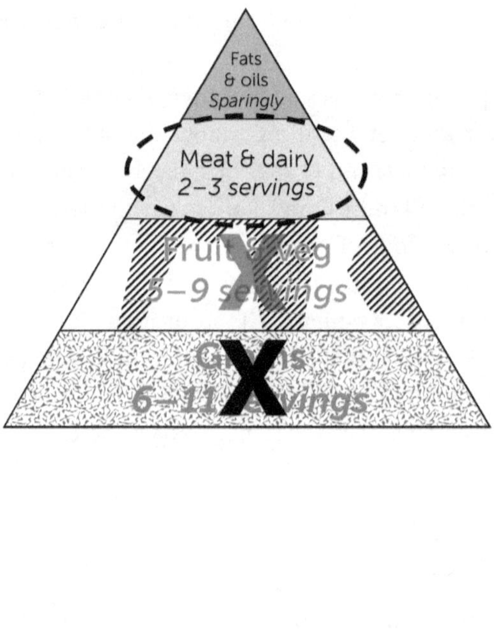

CHAPTER 8

FREE THE RADICALS: THE ANTIOXIDANT AND SUPPLEMENTS MYTHS

Why we don't need antioxidants or supplements on the Sapiens diet

'It's easier to fool people than to convince them that they have been fooled.'
Mark Twain

IN 1999, THE UNITED STATES DEPARTMENT OF Agriculture (USDA) issued a statement that 'antioxidant foods may slow ageing'. At the time, the free-radical theory of ageing proposed that damage to genes and other cell components caused by free radicals was the main cause of ageing. Since antioxidants neutralize free radicals, along with their statement the USDA issued a recommended list of foods that were high in antioxidants.

But thirteen years later, the USDA performed a U-turn. 'The clinical data that one can prevent or treat specific diseases with antioxidant supplements is virtually non-existent,' they said. Antioxidants were 'hype'.

What's going on here?

THE ANTIOXIDANT MYTH

There is a widely held belief that if something promises to be good for you, then the more of it the better. With respect to antioxidants, an effective immune system and protection against cancer are just some of the claims made by the antioxidant-promoting community. We hear that blueberries loaded with anthocyanin will protect against dementia, that catechins in green tea will give skin a healthy glow, and that vitamin C boosts the immune system. No wonder many are fixated on topping up their diet with superfoods and supplements that are overflowing with these supposedly health-enhancing substances.

We aim to dispel the myths that surround supplements and antioxidants – the sort of myths offered up if one types 'antioxidants' into Google. 'An Antioxidant-Rich Detox Supplement That Promotes Longevity' is a typical hit returned from the internet. Longevity, younger-looking skin, reduced risk of cancer, reduced risk of cardiovascular problems, reduced risk of degenerative nervous diseases, better vision in old age; promises trying to convince us that we should have a plentiful supply of antioxidants in our diet. It sounds persuasive; to live healthier, and longer. Indeed, the language surrounding antioxidants makes them sound caring, benevolent, human almost. Antioxidants 'scavenge', they 'fight', they 'mop up', they 'clean'. Antioxidants wage war on our behalf. Who, or rather, what, are they fighting? Those dangerous free radicals, of course. If antioxidants are the heroes in this fight, free radicals are cast as the villains. Boo hiss.

But one doesn't have to scroll down internet search results very far to read quite different stories. Articles on the web from trustworthy sources that cast doubt on the claims made for antioxidants, and the need to add more of them to our intake.

WHAT ARE FREE RADICALS AND ANTIOXIDANTS?

Free radicals are unstable compounds produced during metabolism. In the context of free radicals, what 'unstable' means is that they readily react with substances such as the DNA in our cells, causing a chemical chain reaction, which in turn causes damage and lays the foundation for various diseases. This damaging process is known as oxidation.

Antioxidants, meanwhile, are substances that can stabilize these unstable free radicals. So when antioxidants combine with free radicals, damaging oxidation is avoided. You could say that antioxidants 'fight' free radicals and so prevent or mitigate the damage that they cause.

So where do these free radicals come from? They are natural waste products generated by processes such as digesting food and using our muscles. The ability to turn food into energy depends on a chain reaction inside our cells – free radicals are a by-product of this energy generation. There are also a few unnatural ways of generating free radicals, such as smoking, exposure to pollutants and consuming alcohol. When the level of free radicals gets out of hand, our bodies are put under 'oxidative stress'. Doesn't sound like a good situation – maybe we should do everything we can to increase antioxidant levels and get rid of the nasty free radicals?

But in fact, and take a deep breath, as this may surprise you, our lives would not be possible without free radicals. They are necessary for many normal biological functions, such as controlling breathing (you can breathe out now), the production of red blood cells, the regulation of insulin, and the body's everyday repair, wound healing and regeneration processes.

What's more, free radicals are a crucial part of our immune systems, as they are necessary to attack bacteria and viruses. Indeed, a modest level of free radicals triggers beneficial biological responses, which is believed to protect against later cell damage (this is known as 'hormesis'). Put all this together, and some would go so far as to suggest that free radicals contribute to healthy longevity. Rather like antioxidants, then.

A BALANCING ACT

The way to make sense of these claims for both antioxidants and free radicals is to understand that there is a balance to be struck between the two, a balance that is right for us. Things only go wrong when free radicals and antioxidants are out of balance. But maintaining checks and balances is something our internal biological processes do all the time. For example, blood pH is usually maintained at 7.40. The mechanisms to keep our blood pH stable are incredibly complex. It doesn't matter how acidic or alkaline a diet you eat, the body keeps blood pH at 7.40. The reason for the strict pH control is that if the level veers outside a range of 7.35 to 7.45, the consequences are severe and can be fatal. Biology is indeed very clever – it keeps our blood pH within 0.05 either side of 7.40 regardless of what we do. That's a superb level of control.

Similarly, there are sophisticated mechanisms inside all cells to regulate the balance between free radicals and antioxidants. Millions of years of evolution have led to a perfection of the controls within cells to balance the positive and negative effects between the two. Our bodies are making, using and destroying free radicals all the time. Flooding the body with extraneous antioxidants risks disrupting this natural balance. And even if the level of free radicals should get out of hand and create an imbalance, it is not guaranteed that ingesting more antioxidants is necessarily a benefit. The reason being that there is no certainty that antioxidants will find their way to the right place and neutralize the free radicals quickly enough. A better strategy would be to decrease the production of free radicals in the first place – by avoiding things like tobacco smoke, alcohol, pollutants, and so on. And, surprisingly, carbohydrates should also be on the list of things to avoid (more on carbohydrates shortly).

SOURCES: DIET OR SUPPLEMENTS

But still, isn't it better to be safe rather than sorry, and consume a diet rich in antioxidants, or take antioxidant supplements? In order to understand why that isn't the case, it's important to know something about the dietary sources of antioxidants. Two well-known antioxidants are vitamin C and vitamin E. Other antioxidants include uric acid, ketones, vitamin A, vitamin E, zinc and selenium. Foods containing one or more of these antioxidants include strawberries, blueberries, blackberries, kiwi fruit, broccoli, leafy vegetables, carrots, tomatoes, almonds, Brazil nuts and many more (one doesn't normally see 'liver' or 'offal' in lists of antioxidants, but they should be top of the list). There are other antioxidants, known as phytochemicals, found in green tea, and also polyphenols and flavonoids, which provide a convenient excuse to consume dark chocolate and red wine without guilt.

And then, on top of dietary sources, there are antioxidant supplements, the sorts available at health shops, pharmacies and on websites. Tablets in brown plastic bottles, or white plastic pots; tubs of antioxidant creams; squeezy tubes of antioxidant serums. Some are delivered by droppers onto the skin. There's an antioxidant treatment that claims to be an organic, vegan way to fortify the skin against environmental stress; what's more, it's scientifically proven 'in-vitro' (which means an experiment was carried out in a test tube; hardly scientific proof of its effectiveness when splashed all over).

DO WE NEED TO WORRY ABOUT THEM?

Are antioxidant-rich diets or antioxidant supplements beneficial? It depends on who you believe. But there's evidence that shows that consuming large amounts of fruit and vegetables has no effect on our inherent antioxidant mechanisms. In one randomized controlled study (the best type of study) one group ate an extra 480g of fruit and vegetables every day (equivalent to an extra seven portions/day) for twelve consecutive weeks, and the other group ate three or fewer portions of fruit and vegetables every day. (Compliance in this study was excellent – it was checked by doing blood tests.) To everybody's surprise, no difference was found in antioxidant levels between the two groups. Nor was there any difference in damage to DNA in the two groups. Unsurprisingly, the conclusion was that eating large amounts of fruit and vegetables has no effect on the body's antioxidant levels or damage to DNA.

Other studies are just as negative about the effects of antioxidant supplements. In one trial, 29,000 male smokers were assigned either to a group given an antioxidant supplement (beta-carotene) or to a group taking a placebo. After eight years it was found that there was an 18% increase in the incidence of lung cancer in the antioxidant group. In fact, total mortality went up 8% in the beta-carotene group, owing to lung cancer and heart attacks. The conclusion was that, for male smokers, the beta-carotene antioxidant supplement was harmful.

In another trial 18,000 smokers and ex-smokers received either a supplement (beta-carotene again) or a placebo, and similar results were observed. The group that was given the supplements had an increased incidence of lung cancer, as well as an increased risk of death from lung cancer or cardiovascular disease. The trial was discontinued prior to the planned end date because of the increased mortality risk.

Another investigation looked at the effects of antioxidant supplements on the risk of prostate cancer. This involved 35,000 men who were given either selenium, vitamin E supplements or a placebo. Those on the vitamin E supplement were found to have a 17% increased risk of prostate cancer. The harmful effects of vitamin E started to become clear three years into the study; this trial too was suspended early.

The idea that naturally occurring antioxidants in fruit and vegetables confer health benefits gained traction from several epidemiological studies that linked the consumption of fruit and vegetables with lower incidences of chronic diseases and cancers. However, even though epidemiological studies are useful for developing hypotheses, they are ultimately unable to provide proof of causes (as explained in the chapter on flaws in nutrition science). To the surprise of many researchers, the majority of controlled studies (the good type of study) show that fruit and vegetable consumption has no effect on antioxidant activity in the body, whether it be very large quantities of fruit and vegetables, or just the standard, recommended amounts.

If there are any benefits to eating (at least) five-a-day, improvement in antioxidant status is not one of them. A simple explanation for why epidemiological studies show that people who eat more fruit and vegetables are healthier could be because they eat fewer grain-containing foods or lead healthier lifestyles by engaging in things that are good for their well-being, such as regular exercise, and are less likely to engage in free radical-inducing smoking or excessive alcohol drinking.

THE USDA U-TURN

Now we come to the United States Department of Agriculture (USDA). So it was that in 1999 the USDA declared that 'antioxidant foods may slow ageing', and they published their list of foods with scientifically verified antioxidant levels. Since the 1950s, the 'free radical theory of ageing' proposed that genes and other cell components were damaged by free radicals, and that this damage was the main cause of ageing. The theory was modified in the 1970s to the 'mitochondrial theory of ageing' (mitochondria are the 'powerhouses' of our cells). This suggested that harm to mitochondria and DNA (caused by free radicals) was the principal culprit of ageing. No surprise then that the USDA should endorse antioxidants.

But in 2012, after surveying much evidence, they changed their minds. While acknowledging that diseases can be caused by free radicals, the USDA's view was that 'the clinical data that one can prevent or treat specific diseases with antioxidant supplements is virtually non-existent'. They concluded that the claims about antioxidants were 'hype' that 'far exceeds the evidence of any real value'. They warned that some evidence indicated that supplements presented significant risks, 'including the likelihood of some diseases and interfering with some kinds of medical therapy'. They cautioned that data was routinely being misused by food and dietary supplement manufacturers to promote products, and the same data was misused or misunderstood by consumers seeking to understand their food and dietary supplement options.

The USDA withdrew its public database on the antioxidant benefits from foods, and published an article that made, among others, 'the following important points':

> [There is] mounting evidence that the values indicating antioxidant capacity have no relevance to the effects of specific bioactive compounds, including polyphenols on human health.
>
> There is no evidence that the beneficial effects of polyphenol-rich foods can be attributed to the antioxidant properties of these foods.
>
> The data for antioxidant capacity of foods generated by in vitro (test tube) methods cannot be extrapolated to in vivo (human) effects and the clinical trials to test benefits of dietary antioxidants have produced mixed results.

So forget your antioxidant skin drops.

OTHERS HAVE DOUBTS

Here is the view of Cancer Research UK, available online:

> [We're] concerned about the relentless portrayal of antioxidants as a universal health panacea, in the absence of robust scientific data – partly because there's emerging evidence, including a paper in the journal *Cancer* … that the knock-on effect is that some people undergoing cancer treatment may be taking them in high doses without thinking to tell their doctors … This is worrying, because there have also been suggestions that antioxidants might interfere with treatment, at least in some people.

Cancer Research UK summarize by saying that 'it's likely that a diet containing *some* antioxidants *might* protect against *some* cancers under *some* circumstances as part of *some* diets' (their italics), and that if there were any proven benefits to be gained from antioxidants, trials would have seen them – 'but they haven't'. Cancer Research UK also point out the similar lack of evidence that vitamin supplements, many of which are antioxidants, offer any protection against cancer. For example, they mention a study of 160,000 women that found that vitamin supplements had no effect (although at least 'they didn't do any harm either, as other studies have suggested').

The research quoted by Cancer Research UK is in alignment with the conclusions of a 'study of studies' that was carried out by the Cochrane collaboration. A study of studies (known as a meta-analysis) is only reliable if the individual studies included are themselves reliable. If you take an average of rubbish, you still get rubbish. But the Cochrane study encompassed seventy-eight randomized clinical trials – the optimum sort. This meta-analysis concluded that 'supplements of vitamins A, C and E, beta-carotene and selenium do not improve your health. And the very best trials showed that beta-carotene, vitamin A and vitamin E supplements could possibly increase the risk of dying from various diseases'.

The USDA, Cancer Research UK, Cochrane – none of them is convinced by the hype around antioxidants and antioxidant supplements.

OTHER SUPPLEMENTS AND NUTRIENT DEFICIENCIES

It sounds obvious that if someone is known to have a deficiency of a particular vitamin or mineral, it would be advisable to take a supplement to correct that deficiency. For example, if you are low on iron, it makes sense to take an iron supplement. However, this is not always as straightforward as it seems. There have, in fact, been some tragic cases involving supplements taken to address deficiencies. Iron and vitamin D are excellent examples to illustrate the pros and cons of using supplements in these situations.

There are numerous biological processes in the human body for which iron is needed, such as making red blood cells, maintaining the proper functioning of the immune system and preventing neurological problems. Some of the problems caused by a deficiency of iron include anaemia, increased risk of infections and cognitive impairment. Iron deficiency is a global and a serious problem, especially for those living in the developing world, where it is thought to contribute to the high burden of infection and cognitive impairment in infants and children. Anaemia affects a massive one-third of the world's population and is particularly prevalent in women and young children. Iron deficiency is responsible for 800,000 deaths every year.

However, when population-wide iron supplements have been introduced in developing countries there has been a significant rise in malaria, diarrhoea, tuberculosis and other serious infections in the groups treated with iron compared to the groups given a placebo. It has been suggested that iron deficiency in certain groups of people may have come about as an evolutionary advantage (iron deficiency is known to offer protection against malaria and in certain other infections). Furthermore, in children who have adequate iron, giving supplements impairs their growth – and this realization resulted in the World Health Organization revising its guidelines for iron supplementation. This makes me wonder whether the routine, artificial fortification of baby formula milk with iron, over and above the amounts of iron present in breast milk, might be a reason why childhood infections are so common even in infants in developed countries?

BRING ME SUNSHINE: VITAMIN D

Now we come to one of the most common vitamin supplements: vitamin D. Even though it might sound like a good idea to take large doses of vitamin D supplements to correct a shortfall, the intricate interactions between different nutrients means that you might not get the benefits of vitamin D if you have a deficiency of another nutrient as well.

Vitamin D is commonly known as the 'sunshine vitamin'. Vitamin D is either made in the human body when the sun hits the skin (for which cholesterol is needed), or is obtained directly from one's diet. And what does vitamin D do? There is no disputing the importance of vitamin D; it has long been known to be essential for bone health. But it is now known that besides being needed for our bones, it's also involved in many other biological processes in the body. In fact, there are vitamin D receptors in virtually every cell in the body, including in the brain. These receptors must be there for a good reason and may explain why the health problems that are associated with a low level are legion – such as heart attacks, strokes, certain cancers, tuberculosis, cognitive decline, depression, allergies, schizophrenia, multiple sclerosis and other autoimmune diseases. It is worth noting here that a low vitamin D is *associated* with these diseases – it is not known whether a deficiency actually causes them.

Nevertheless, all these factors have contributed to the high-profile news coverage of the benefits of vitamin D and the miraculous tonic-like claims made for vitamin D supplementation.

A lack of this vitamin undoubtedly causes problems – in particular, rickets, the bone disease that was commonplace in the children of yesteryear. Fortunately, the almost universal recommendations of cod liver oil from the 1920s onwards put paid to rickets in most developed countries (cod liver oil is an excellent source of vitamin D). By the 1940s rickets had almost disappeared in the UK. Then the thinking went that if vitamin D can cure rickets, surely it can also cure the other diseases that a lack of vitamin D is associated with? This assumption is backed up by the fact that most *observational* studies show that vitamin D supplementation is of benefit. However, most *controlled* trials do not support the idea that vitamin D supplementation is of benefit, other than in preventing rickets (the evidence from controlled trials being much more credible than the evidence from observational studies).

Sunshine and supplements aren't the only sources of vitamin D. This is evident from looking at certain populations that live in the Arctic, where there isn't enough sunshine to make vitamin D for six months of the year. Surprisingly, several studies have found that some of these Arctic residents have higher levels of vitamin D than those living in tropical regions that are bathed in sunshine.

The fact that it is possible to have adequate levels of vitamin D even when there isn't adequate sunshine has been verified in several studies. One study done on people living close to the Arctic Circle in Sweden found that vitamin D levels at the end of winter were among the highest in the world. Another trial in Siberia showed that indigenous people who consume their traditional animal-food rich diet had normal vitamin D levels, even at the end of winter, whereas the indigenous people who lived in towns and ate a 'Westernized' diet had sub-optimal vitamin D levels – and this despite both groups having similar levels of sun exposure. The common thread in all the studies with similar findings is that the higher the level of meat and fish intake, the higher the blood level of vitamin D. It's also been found that the higher the carbohydrate intake, the lower the blood level of vitamin D is. Could the high-carb and low-fat Western diet of the last fifty years be partly to blame for the global vitamin D deficiency?

INTERPLAY BETWEEN VITAMIN D WITH MAGNESIUM, VITAMIN K2 AND CALCIUM

There are yet other reasons why vitamin D supplements have not lived up to their promise – it may be that other deficiencies (such as magnesium and vitamin K2) affect the functioning of vitamin D. For example, magnesium is needed to activate vitamin D, and unless it is activated, vitamin D is ineffective. In fact, vitamin D-resistant rickets can be caused by having a low magnesium level. In this situation, the rickets is reversed by correcting both the magnesium and vitamin D deficiencies at the same time.

Vitamin K2 has a similar story. But what exactly is vitamin K2? You could be forgiven for not having heard of it – most doctors and nutritionists haven't heard of it either. Owing to its relatively recent discovery, vitamin K2 has come to the fore only in the last few years. So how exactly are vitamin D and K2 connected? Through calcium. One of the effects of vitamin D is to increase calcium absorption from food in the gut, which then makes calcium levels in the blood go up. If this extra calcium ends up in our bones, that's a good thing. But if the extra calcium ends up in our arteries, that's not such a good thing. Why not? Because it leads to the formation of plaques in arteries. The relevance of plaque build-up in arteries that supply the heart (the coronary arteries) is that this increases the risk of having a heart attack. And if the plaques build up in the arteries that supply the brain, then this increases the risk of a stroke. Clearly, calcium building up in arteries isn't a good idea.

This is where vitamin K2 comes in. Vitamin K2 is the only mechanism that our bodies have of clearing calcium away from the insides of arteries. The upshot of all this is that if one is taking vitamin D supplements while vitamin K2 levels are low, then the extra calcium ends up causing harm by getting deposited in arteries.

VITAMIN D SUPPLEMENTS AND HEART ATTACKS

In sum, vitamin D and vitamin K2 work as a team. Unfortunately, people who are low in vitamin D are usually also low in vitamin K2. The net effect of increasing vitamin D levels when vitamin K2 is also low is to cause calcium to end up in places where it could cause serious problems. The UK government advice is that everyone (including babies and children) should consider taking a vitamin D supplement in the autumn and winter. If you are taking vitamin D supplements when you also have a vitamin K2 deficiency, there is a good chance you might be inadvertently increasing your risk for a heart attack or stroke.

INTERPLAY BETWEEN HIGH-CARBOHYDRATE DIETS WITH ANTIOXIDANTS AND FREE RADICALS

Besides vitamins and minerals, there's another factor that simply can't be excluded from the antioxidant and free-radical equation. This is that high-carbohydrate diets generate more free radicals than low-carbohydrate diets. How do high-carbohydrate diets increase free-radical production? Via higher blood glucose and insulin levels. And which is why lowering carbohydrate intake increases the natural antioxidant capacity in the human body. Although, apart from biochemists, not many know this. And high glucose and insulin levels have many other negative knock-on effects, such as lower magnesium levels in the body (because more magnesium is excreted) and an increased requirement for vitamin C (because vitamin C has difficulty getting into cells, as explained in the previous chapter).

Rather than relying on supplements or on fruit and vegetables to mop up excess free radicals, it makes more sense to generate fewer free radicals in the first place. Because low-carbohydrate diets generate fewer free radicals, an easy solution is to eat fewer carbohydrates. The most effective way to achieve a healthy balance of antioxidants and free radicals naturally is to consume as low a carbohydrate diet as one can. Very low-carbohydrate diets, which are also ketogenic, have been shown to vastly increase the antioxidants that are normally produced by the human body. And out of all the different types of ketogenic diets, the Sapiens diet is the best (details of these diets will be covered in chapters 11 and 14).

The bottom line is that our diet should contain safe and adequate levels of all essential nutrients, including vitamins and minerals. When this is the case, the human body will make all the antioxidants that are needed and there won't be a problem of excess free radicals. Furthermore, nutrients usually act in a coordinated way in the human body and there may be interactions between nutrients of which we are not yet aware – which is why taking a supplement of one or two nutrients in isolation may create unintended problems. Supplements are not smarties to be handed out to all and sundry. Rather than correcting a deficiency with a supplement, we should focus on correcting the inadequacies in the diet that led to the deficiency. Provided that we eat as we were meant to, we can trust our physiology to carry out all the balances and checks that are so elegantly orchestrated within.

SUMMARY

Overall, there is a lack of reliable evidence that a high intake of antioxidants reduces the risk of disease. In most cases, studies show that antioxidants either have no benefit, or that they have a detrimental effect, or that results are so unclear that no significant conclusion can be drawn one way or another. Providing that satisfactory quantities of vitamins, minerals and other micronutrients have been consumed in our diet, then taking in extra antioxidants from food or from pills is unnecessary.

Whether with respect to antioxidants in particular, or supplements in general, it's difficult to avoid the conclusion that the modern dietary industry, with its associated advertising and marketing, is driving us down the wrong path. This path endorses diets, such as vegetarianism or veganism, that are lacking in the balanced set of vitamins and nutrients we need, or that upset our body's natural ability to maintain an equilibrium by itself. The industry encourages us to resort to pills and potions to fill in the real or imagined gaps in our nutrition, with no acknowledgment of the possible harms that can result.

We believe that rather than take this route, the better direction would be to adopt – readopt – the diet that evolution selected for us, and which, through the law of survival of the fittest, must be the optimum way to eat.

Fruit and vegetables are partially removed from the pyramid. Now we move up to the next layer and find out why meat deserves to be kept. And doing that involves answering a key question. Why do vegans have smaller brains?

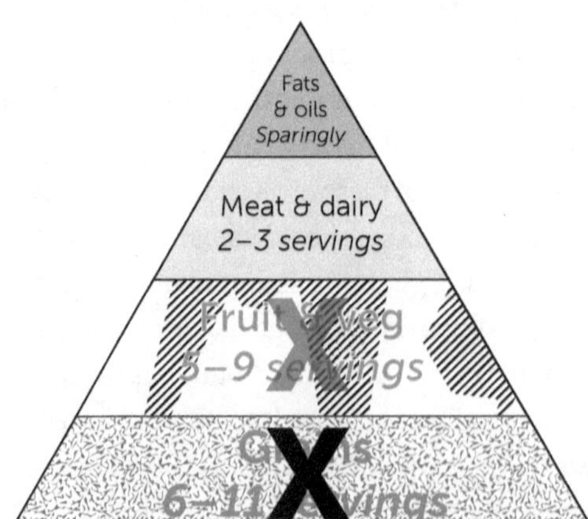

CHAPTER 9

WHY VEGANS HAVE SMALLER BRAINS

The need to prioritize animal-sourced foods in the pyramid

'There is no scientific study more vital to man than the study of his own brain. Our entire view of the universe depends on it.'

Francis Crick

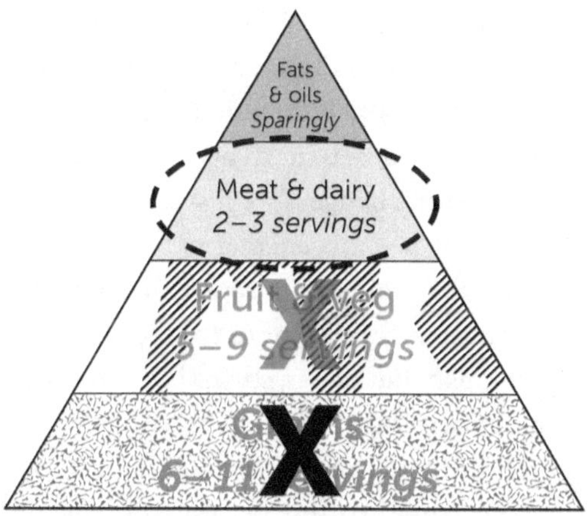

WHY VEGANS HAVE SMALLER BRAINS: THE BLUNT title for this chapter, indeed this book, has simple justification. That's what brain scans show.

There's a vitamin that's crucial for the human nervous system and brain. It's vitamin B12. *And yet no plant food contains it.* Neither five servings a day nor fifty-five servings a day of fruit and vegetables, no matter how varied, no matter how many different colours they are, nor any cereals, grains, nuts or pulses, will provide the body with *any* B12.

In this chapter we'll look at the issues of vitamin B12, brain shrinkage, the risk factors for brain atrophy and dementia, the particular hazards for the very young and the more elderly, and how these link to one's choice of diet. But there is more to this layer of the pyramid than just vitamin B12. This is the layer that, together with the fat layer, gives you all the vitamins and minerals that you need.

BRAIN ATROPHY

Brain shrinkage – also known as 'brain atrophy' – is a degenerative process once thought to start in middle age. According to the latest research, however, brain degeneration starts much earlier: the human brain is at its largest in our twenties, from which point it is all downhill. We lose on average 1% of our brain mass every year between the ages of twenty and fifty. This figure doubles to 2% every year from the age of fifty onwards and accelerates even further as we enter our final years. As we age, our brains, literally, shrink. Not the best of news for anyone over the age of twenty, but hey-ho.

Ageing appears to be the biggest risk factor for brain atrophy; sadly, there's not much we can do about this risk. There is, however, a little bit of good news. A high alcohol intake, chronically high blood-sugar levels, being overweight and a diet that is low in vitamin B12 are all also strongly associated with brain shrinkage. Things that we can change if we just try.

The reason for devoting a whole chapter to vitamin B12 is because this is one of the few treatable causes of brain atrophy, in babies and in older people. Why is vitamin B12 particularly relevant to this book? *Because vitamin B12 is not present in any plant food.* But does a shrinking brain matter? Can we be as mentally astute with a smaller brain? That depends.

Most of us notice an increasing forgetfulness as we age. It may take us longer to recall someone's name, to think of a word or remember why we walked into a room. When do we draw the line and start to think of these memory lapses as something serious? The stage in between the cognitive decline of normal ageing and full-blown dementia is termed 'mild cognitive impairment' and is more than just being a bit absent-minded. The significance of mild cognitive impairment lies in the fact that it significantly increases the risk of developing dementia. Mild cognitive impairment and dementia are, in fact, part of the same spectrum – cognitive impairment usually precedes dementia by several years. Every year, about 10% of people with mild cognitive impairment go on to develop full-blown dementia. So, what exactly causes dementia?

Any injury to the brain – from contact sports to concussion, and from strokes to syphilis. However, the leading cause of dementia is Alzheimer's disease. But as Franz Kafka famously said, 'Just because your doctor has a name for your condition, doesn't mean he knows what it is.' This is exactly the case with Alzheimer's – it isn't known what causes it. However, there is one cause of dementia that we do know about and that is easily reversible – a deficiency in vitamin B12.

DEMENTIA MATTERS

Dementia is now the leading cause of death in the UK. Its impact is greater than that of heart attacks, strokes or lung cancer. Worldwide, 55 million people suffer from dementia, and this is predicted to triple by 2050. Surprisingly, initial brain size appears to be irrelevant as far as dementia is concerned. What pushes us down the path towards a debilitating loss of brain function is the rate at which our brains atrophy: the more rapidly it occurs, the greater the risk of dementia. It would, in fact, be better to start off with a small brain that doesn't shrink as you get older than to start off with a large brain that shrinks a lot. The good news is that brain atrophy can sometimes be significantly slowed down. And yes, we can be mentally astute with a smaller brain, so long as we can slow down the rate of shrinking.

B12: A BRAIN CHANGER

While all vitamins are important, vitamin B12 is particularly important, and especially so for our brains. Trouble is, diagnosing a low vitamin B12 level is not straightforward. There is a blood test to check vitamin B12 levels, but often by the time this test shows up as low, damage to the body and brain has often already been done. Ideally, three other blood tests should be done in conjunction with the vitamin B12 blood test to give an accurate measure (the other tests are homocysteine, holotranscobalamin and methylmalonic acid – quite a mouthful, as is often the case with medical tests). Unfortunately, not all doctors know about these other blood tests and frequently only the vitamin B12 blood test is checked, which is why a low vitamin B12 level is often missed. Because improving the level of vitamin B12 slows the rate of cognitive decline and brain atrophy, it is worth doing the three extra blood tests when checking for a vitamin B12 deficiency. As such these tests are one of the simplest and cheapest ways of diagnosing and then remedying one of the few treatable types of dementia.

A FEW BASICS ON THE BIOLOGY OF VITAMIN B12

There are yet other serious and sometimes fatal consequences of a low vitamin B12. Many crucial functions in the human body depend on vitamin B12, such as the synthesis of DNA, RNA, red blood cells and myelin (the fatty layer that surrounds and insulates nerves). Myelin is especially important when the brain is developing before birth and in the first year of life.

The brain and the body are hit hard by a deficiency of vitamin B12 throughout life. Besides brain atrophy and dementia, there is a long list of medical conditions that a deficiency of vitamin B12 causes. Fatigue, anaemia, and serious neurological and psychiatric problems such as depression, psychosis, mania and behavioural changes are some of them. The list goes on. In babies, a severe deficiency can cause brain atrophy; in infants, it can cause involuntary tremors, irritability and anorexia (which can last well into adulthood); in young children, it can stunt intellectual development; and in adults, a low dietary intake of vitamin B12 can cause degeneration of the brain and spinal cord. Phew.

Worryingly, these conditions frequently cause no symptoms for many years; the full-blown disease often only becomes evident in the final, and sometimes fatal, stages of vitamin B12 deficiency. Despite vitamin B12 deficiency first being recognized over a century ago, it is, to this day, difficult to correctly diagnose and treat. This is partly because the symptoms of vitamin B12 deficiency are diverse and difficult to pick up, plus the fact that all the blood tests needed to diagnose a deficiency are rarely carried out.

It is also unclear why some people with low vitamin B12 develop brain atrophy while others don't display any symptoms at all. It might be owing to a combination of the depth of the deficiency together with the lack of other key nutrients in the diet. More on the other nutrients shortly.

BRAIN ATROPHY IN BABIES

It is easy to see that a deficiency of vitamin B12 causes problems with the brain in adulthood and later life – I learned about all these issues in medical school some forty years ago. What I didn't know then is that a deficiency of vitamin B12 can also cause brain atrophy and other serious problems in babies. What has surprised me the most at the outset of this book is that breastfed babies of vegan or vegetarian mothers are at a particularly high risk of vitamin B12 deficiency. How did I find out that low levels of vitamin B12 might cause brain atrophy in babies?

From studies from all four corners of the earth. What the studies show is that there is a dramatic reversal of brain atrophy when these breastfed babies are diagnosed and treated early with vitamin B12. So it wasn't all bad news.

Below are scans from some of these studies. The scan on the left shows the atrophied brain of a baby suffering from severe vitamin B12 deficiency. In this scan, the brain matter (the white and light grey colours) has shrunk, and deep gaps have appeared around and within the brain (the dark grey and black). The gap in the centre of the scan is an abnormally enlarged ventricle because the overall volume of brain matter is lower than it should be.

The scan on the right is normal – there is no brain atrophy eight months after the B12 deficiency was corrected. There is now a normal volume of brain matter, which also makes the ventricle return to a normal size.

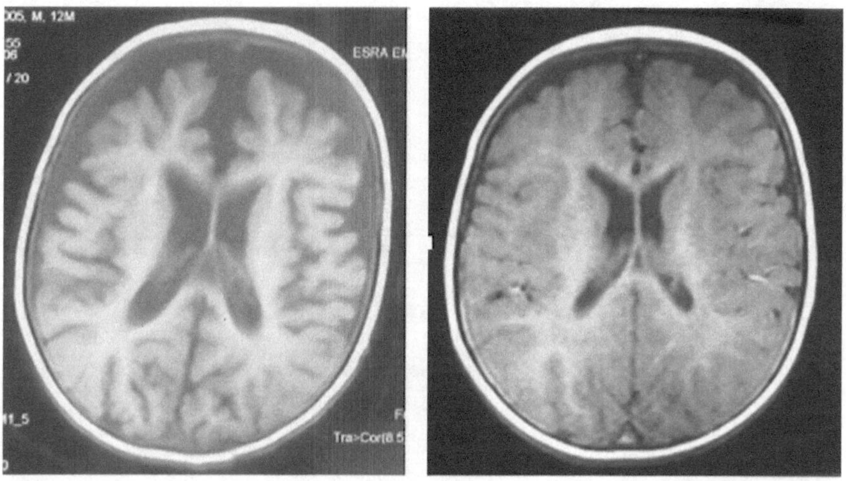

Source: Reproduced from OZYUREK, H., et al., Vitamin B12 deficiency as a treatable cause of severe brain atrophy. *Neurology Asia*, 2021. 26(1).

Most of these tragic cases of brain atrophy have been in exclusively breastfed babies in poverty-stricken areas in the developing world. Often a lack of animal foods in the diets of these mothers is not a lifestyle choice – these women are nearly always from economically deprived backgrounds and cannot afford to eat nutritious animal foods. But, even though poverty is very often the cause of a nutrient-poor diet in the developing world, there are an increasing number of cases in countries where almost everyone can afford to include meat in their diet. The main cause of this phenomenon is the rise in veganism and vegetarianism. It is, indeed, unfortunate that breastfeeding mothers who can afford to include animal foods in their diets choose to consume so little of them, or worse, none at all. Vitamin B12 treatment always needs to be administered promptly to prevent permanent brain damage in babies, as was the case in the scans below from a baby in Switzerland.

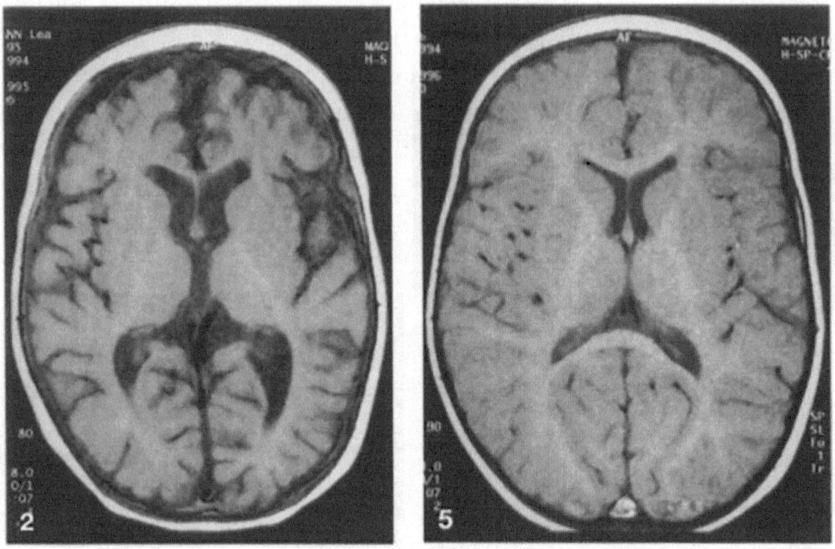

Brain scans of a baby with severe brain atrophy on the left and complete reversal of the atrophy after treatment with vitamin B12 on the right. Source: Reproduced from Lövblad, K. O. et al. Retardation of myelination due to dietary vitamin B 12 deficiency: cranial MRI findings. *Pediatric Radiology*, 1997. 27(2): pp. 155-158.

My heart sinks when I hear of any young woman who is following a plant-based diet and I pray that she reads this book before embarking on starting a family.

IS BEING A BREASTFEEDING VEGAN MOTHER A JUSTIFIABLE RISK FOR HER BABY?

It would be correct to say that not all breastfed babies of vegan mothers go on to develop brain atrophy. Why it is that some who are deficient in vitamin B12 develop brain atrophy and others don't isn't completely clear. As with brain atrophy in the elderly it may be the extent of vitamin B12 deficiency, or deficiencies of other nutrients in combination with a deficiency of vitamin B12, or both that matter. The good news is that there is a call to introduce screening for vitamin B12 deficiency into the antenatal and newborn screening programmes for mothers and their infants who are at risk of vitamin B12 deficiency (which is all vegetarians and vegans).

Besides causing severe problems in babies, a deficiency of vitamin B12 is also problematic for growing children and adolescents. A child's cognitive potential is permanently compromised if there is an inadequate intake before the age of five; deficiency in infants is a global problem and a matter of urgency. To fulfil one's cognitive potential, it is essential to have adequate amounts of certain key vitamins and minerals in infancy and childhood. If you miss out on vitamin B12 at these critical growth stages, there's no catching up.

BRAIN ATROPHY IN OLDER PEOPLE

Can you remember what you had for breakfast, lunch and dinner yesterday and the day before yesterday? No, this isn't a test to see whether you might have brain atrophy. This is to show that most of us think that this sort of information isn't important enough to prioritize. Yet, most nutrition studies are based on people remembering what they have eaten. Besides not remembering correctly, it is well recognized that many of us don't tell the full truth in diet questionnaires, partly because we might not want to think about, let alone see in black and white, that we may be eating unhealthily. That is why it is notoriously difficult to draw meaningful conclusions from evidence derived from nutrition studies.

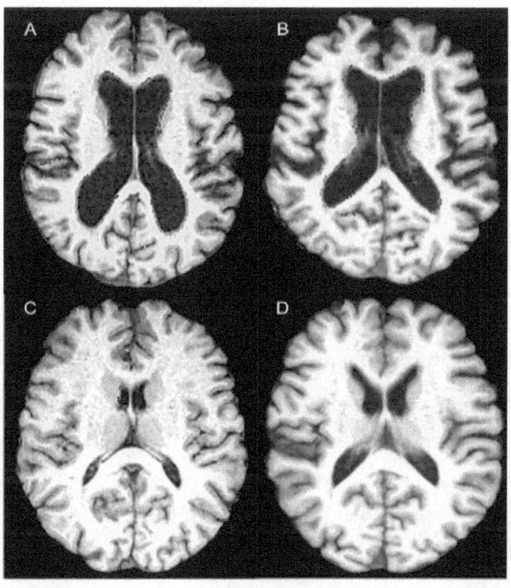

Brain sizes in subjects with high and low vitamin B12 levels. Subjects A and B had low vitamin B12 levels and their scans show severe brain atrophy. Subjects C and D had high vitamin B12 levels and their scans show normal brain size.

On the other hand, nutrition studies that use objective measures such as brain scans and blood tests are considerably more reliable than the sorts of studies that are based only on food diaries. One such reliable study was carried out on a group of middle-aged and elderly people by the University of Oxford. In this study, the researchers tested all four blood markers of vitamin B12 and also performed brain scans over a five-year period. The findings were extraordinary. The brains of those with the lowest blood levels of vitamin B12 were six times more likely to have shrunk over the five years than those with the highest levels (see figure on p. 129). Of particular relevance here is that the lower the amount of meat in a person's diet, the lower their level of vitamin B12, a fact that is true for people across all age groups across the world. Unsurprisingly, vegans have the lowest levels of vitamin B12, with vegetarians not far behind. According to the lead author of this Oxford study, adjusting our diets to consume more vitamin B12 may prevent brain shrinkage.

This Oxford study isn't the only one to show that low blood vitamin B12 levels correlate with a higher risk of brain atrophy on a brain scan. There are others, such as a Swedish study in which brain scans were done on nearly 300 dementia-free people over the course of eight years. They too checked all the markers of vitamin B12. And their findings were the same as the Oxford study – those with the lowest levels of vitamin B12 showed the greatest degree of brain atrophy. The Swedish researchers concluded that low B12 levels are related to 'accelerated ageing of the brain'. As always, there is a bit of good news. Improving your vitamin B12 levels usually slows down brain atrophy. But it isn't guaranteed. And there is one particular situation in which boosting vitamin B12 levels doesn't necessarily reduce the risk of future problems.

COGNITION-BOOSTING EFFECTS OF VITAMIN B12 – ONLY IF THERE IS ENOUGH DHA

Supplementing with B vitamins has been shown to slow brain atrophy by 40%, but only when blood levels of DHA (a type of omega fat) are high. (Supplementing with only vitamin B12 has absolutely no effect on brain atrophy if blood DHA levels are low.) Similarly, for omega-3 fats to do their job on improving cognition, one needs to have good levels of vitamin B12. We need to have adequate levels of many nutrients together for vitamin B12 to do its job properly.

At the risk of sounding like a stuck record, all the essential nutrients that are needed for brain growth and brain function are best obtained from animal foods. We will cover more on DHA in the next chapter and learn that, like vitamin B12, it is only available in animal foods and certain types of seaweed. The bottom line is that you need to have good amounts of both vitamin B12 and omega-3 fats together to reap their benefits.

HOW LOW CAN YOU GO

Rather worryingly, cognition is associated with levels of vitamin B12 across the entire range – a vitamin B12 level at the lower end is also associated with an increased risk of brain atrophy. This has led to calls for the acceptable range for vitamin B12 levels to be revised and for the current lower limit to be raised. To whatever extent we fail to get an adequate supply of vitamin B12, we risk suffering its effects. Ideally, one should maintain a *good* level of vitamin B12, not just an *adequate* level.

VEGAN DIETS, ALGAE, FUNGI AND VITAMIN B12

Seeing as there is no vitamin B12 in plant foods, one would expect *all* vegans to have low vitamin B12 levels. One of the reasons why this is not the case is because stores of vitamin B12 in the human body last between two and five years after vitamin B12 has been withdrawn from the diet, plus some vegans are astute about taking supplements.

Luckily for vegetarians and vegans there are some foods which are neither plants nor animals that contain vitamin B12 – algae and fungi. There is one particular type of seaweed and certain types of mushroom that contain vitamin B12. The seaweed that does have appreciable amounts of vitamin B12 is called 'purple laver' (also known as nori – which is the type of seaweed often used to wrap sushi). So it is possible to obtain some vitamin B12 from non-animal sources.

PSEUDO-VITAMINS

But a word of warning about spirulina (an algae). Spirulina has ample amounts of vitamin B12. However, what it contains are vitamin B12 analogues ('pseudo-vitamins') rather than the true vitamin. These pseudo-vitamins so closely resemble vitamin B12 that, when a blood test is performed, the pseudo-vitamins will be picked up as being actual vitamin B12 because the blood-testing machines cannot differentiate between them. A false sense of security about vitamin B12 levels may be the result.

Some vegans are fully aware of the risks of being deficient in nutrients and are fastidious about taking the right supplements to cover their shortfall. Others know about these algae and fungi. Yet, most studies show that long-term vegetarians and vegans are deficient in vitamin B12.

DRUGS THAT LOWER VITAMIN B12

Besides having a diet that is low in vitamin B12, there are certain drugs that impair its absorption and lead to low levels. The most well-known drugs are omeprazole and ranitidine which are used for reflux and indigestion, and metformin, which is used in the treatment of diabetes. Unfortunately, one of the world's most widely consumed drugs – alcohol – also reduces the absorption of this vital nutrient.

PLANT-BASED DIETS, FORTIFYING FOODS AND VITAMIN B12

Besides diet, the other options to ensure adequate levels of vitamins and minerals are to add them to food artificially (often described as 'fortifying'), taking supplements or, in the case of vitamin B12, to have injections of it. However, food fortification doesn't always guarantee normal levels of vitamin B12 or, indeed, of other vitamins and minerals. As far as vitamin B12 goes, this is likely because higher doses of vitamin B12 than the amounts currently used in fortified foods may be needed in vegetarians and vegans. According to experts, the current UK recommended intake for B12 does not adequately consider the vitamin B12 deficit in those who choose to follow plant-based diets. Of all those who follow plant-based diets, vegans in particular need to take B12 supplements regularly. But rather than taking high doses infrequently, a better strategy is to take smaller doses more frequently; this maximizes the absorption of B12 from supplements and fortified foods. And ideally, all vegetarians and vegans should get their blood B12 levels tested regularly.

Why does all the evidence suggest that the best way to obtain adequate dietary vitamin B12 is to consume animal foods? This may be because other micronutrients are present in animal foods that are necessary for B12 to work effectively, or it may be that there are as-yet undiscovered benefits of the so-called 'food matrix' (the way in which the nutrient is held within the substance of the food) that cannot be replicated in a pill. Or as we saw in the previous chapter, just as magnesium is necessary for vitamin D to function properly, there are likely to be many, many other such interactions and vitamin and mineral inter-dependencies in our bodies. And of course, there is always the possibility that there are yet other vital nutrients that we don't know about.

VITAMIN B12 ACROSS THE WORLD

Low vitamin B12 levels are a global problem: between 30% and 60% of certain population groups have a deficiency, with the situation being particularly dire in pregnant women, elderly people and the developing world. A shocking 74% of the population of India is thought to have insufficient vitamin B12. It is widely recognized that meat is a nutrient-dense food combining high levels of vitamins and minerals with exceptionally good bioavailability. It is also well established that a cereal-predominant diet is associated with malnourishment. International aid organizations know full well that animal-sourced foods, including red meat, are the most nutritious of all foods, and these are the very same foods that are routinely recommended in the developing world for pregnant women, infants, children and adolescents. But this advice appears to stop on reaching adulthood, when suddenly consuming red meat becomes hazardous.

Several European countries agree that a vegan diet is unsuitable for infants and pregnant or breastfeeding women and have taken steps to protect children from the malnourishment that results with a vegan diet. This has been in response to several tragic cases of infants dying as a result of being fed a vegan diet.

The increasing trend towards veganism has led the Royal Academy of Medicine of Belgium to issue a legal opinion to safeguard children from vegan parents. The Belgian authorities concluded that the serious nutritional deficiencies that a vegan diet induces were a reason – on ethical grounds – to take legal steps to protect children. In essence, they have prohibited infants, children, adolescents, and pregnant and breastfeeding women from following a vegan diet. Vegan parents in Belgium can, in fact, wind up with a two-year prison sentence. The Belgians don't mince their words: they claim that, because it is common knowledge that a vegan diet can kill, imposition of this diet on children legally qualifies as 'non-assistance to a person in danger'.

Study after study shows the dramatic benefits of animal foods on health, height and cognition in infancy and adolescence. Fortunately, the United Nations and the World Health Organization (WHO) recognize the importance of animal foods and essential micronutrients, especially while in the womb and during infancy. WHO has therefore prioritized the supply of brain-essential micronutrients – in animal-sourced foods and in supplements – to pregnant women, breastfeeding women, babies and children under the age of five in the developing world.

HIDDEN HUNGER

Nutrition is about much more than just getting the right number of calories. 'Hidden hunger' is thought to affect 2 billion people worldwide, predominantly in the developing world, but it does occur in Western countries as well. This 'hunger' occurs when the food in a person's diet is of such a low quality that it doesn't supply adequate amounts of vitamins and minerals. 'Hidden' because the deficiency is masked by the fact that those with hidden hunger are not underweight or may even be obese. Diets that cause hidden hunger have more than enough calories, but do not have enough vitamins and minerals.

BRAIN VS BRAWN

The stereotype of a meat eater, perpetuated by the media, is of a weight-lifting, vein-bulging body-builder, putting away steaks and egg whites to build muscle and brute strength. But compared to many animals, we are feeble – we didn't grow muscles like many of them did, so we are relatively puny compared to wild animals of comparative size. Our species preference for animal-sourced foods resulted not in brawny brutes but in brainy beings. Eating meat during our evolution didn't by any means make us the strongest of all the species – it made us the cleverest.

It is worth pointing out here that protein is not the issue with plant-based diets – it is easy to get all the protein that one needs from plants. It is vitamins and minerals that are the issue. Besides vitamin B12, some of the other brain-essential micronutrients include zinc, iron, iodine, selenium, choline, DHA and ARA (more on DHA and ARA in the next chapter). Most people think of meat just as a good source of protein, but it is much more than that. For optimal brain growth and the maintenance of normal brain function, one needs micronutrients – and it is meat, offal, eggs and fish that are ideally placed to deliver them.

CONCLUSION

The human brain is particularly vulnerable to a lack of nutrients, and the risks that following a vegan diet pose to brain health are considerable. Conversely, encouraging the consumption of foods rich in brain-specific nutrients would much improve the brain capacity of future generations. In 2016, neurological disorders, including dementia, were collectively the leading cause of disability and the second highest cause of death globally. This is set to rise even further.

Five-a-day of fruit and vegetables will not supply any of us with any vitamin B12. Neither will ten portions of fruit and vegetables, nor twenty, because B12 is not present in these foods. However, all animal foods and a few algae and fungi contain some B12. But out of all these foods, animal foods and offal are the best.

This vitamin is essential for good health. It's important for brain development, and has been shown to ward off or reverse brain atrophy. It's vital during pregnancy, in early phases of child development, in adulthood and in older age. This is why we say that those following a vegan diet, in rejecting animal foods, risk suffering the consequences of vitamin B12 deficiency.

And one of those consequences is brain shrinkage. We need to keep animal-sourced foods in our diet.

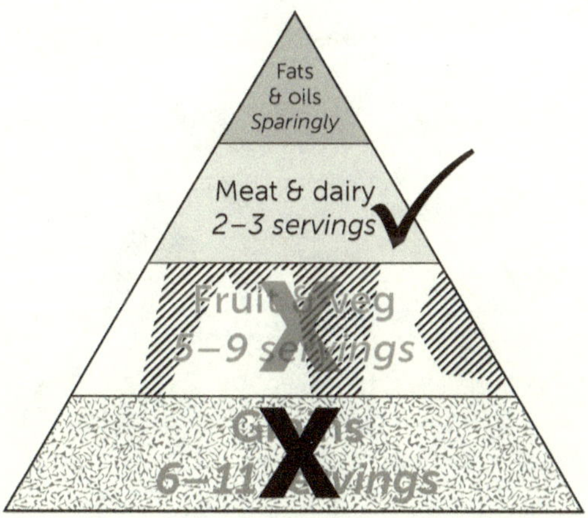

CHAPTER 10

JUST STOP VEGETABLE OIL

The pros and cons of animal fats and vegetable oils

'Civilized man is the only animal clever enough to manufacture its own food, and the only animal stupid enough to eat it.'

Barry Groves (author of *Eat Fat Get Thin*)

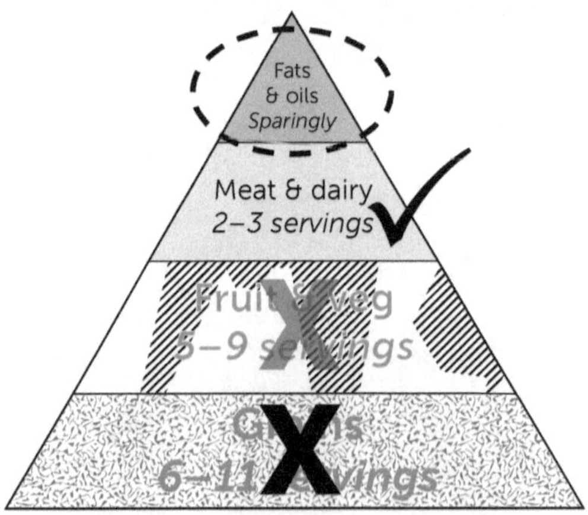

THERE'S A TYPE OF INGREDIENT IN MANY foods that is viewed as heart healthy, vegan, sugar-free and gluten-free. This magical ingredient has become ubiquitous in recipe books and in processed foods. What is this mysterious food? There might be a few trendy contenders – avocados, oats and quinoa may spring to mind – but the one we're thinking of is vegetable oil. But there is a not so well-known dark side to vegetable oils; this sinister food warrants a whole chapter to explain why we shouldn't be consuming it. We will also explain why animal fats are not only healthy but are *essential* for good health.

As far as nutrition goes, the fats that we need to know about are animal fats and plant fats. Calling plant fats 'vegetable oils' inherently makes them sound healthy – associating them with a food type that is generally believed, rightly or wrongly, to be good for you. Moreover, whereas animal fats have been vilified, vegetable oils have been bestowed with a shimmering glow. Surely a bottle of golden yellow liquid such as sunflower oil or olive oil must be healthier than a jar of solid, artery-clogging dripping? Never mind that there isn't a single vegetable oil that actually comes from a vegetable. Never mind that vegetable oils are one of the most processed foods. And never mind that animal fats hold the secret to human intelligence.

The idea that eating animal fat clogs up our arteries sounds plausible. But fat in our food doesn't clog our arteries. The body is more complex and cleverer than simply taking what we eat and injecting it straight into the bloodstream. Congealing animal fat in an egg sandwich (watch the beginning of *Withnail and I*) does not become fat globules solidifying in your bloodstream. Just as the bread in the sarnie doesn't end up as crumbs floating around your arteries. But this sort of misconception helps perpetuate the myth that liquid vegetable oils are healthier than solid animal fats.

VEGETABLE OILS DON'T COME FROM VEGETABLES

The first big change in the human diet occurred 10,000 years ago with the addition of grains. And another big change, one that happened in the last 100 years, has been the introduction of vegetable oils.

The term 'vegetable oil' is a bit of a misnomer – it is an umbrella term for any oil that has been produced from a plant. Most so-called vegetable oils are actually 'seed oils' that have been extracted from plant seeds, such as rapeseed or sunflower seed. And oils that have been extracted from the flesh of fruits, such as olives, coconuts and avocados, are also commonly known as vegetable oils – even though fruit oils would be a more accurate name. None of the oils known as 'vegetable oils' is actually made from a vegetable.

Animal fats, on the other hand, are what they say. Animal fats come from animals – either directly as beef dripping, or indirectly as butter. If we were to rank the healthiness of these various fats, top of the list would be pure animal fats (such as beef dripping), followed by clarified butter (ghee) and butter; fruit oils and then seed oils would be at the very bottom.

Misnomers aside, why are so-called vegetable oils so bad? As with all biological systems, it's good to look at things in an evolutionary context. If we look at the role that animal fats have played in our evolution, we find that they have been part of our diet for a very, very long time. As we learned in chapter 2, eating the fatty parts of animals such as brain and bone marrow was pivotal in the evolution of the human brain. But what role have seed oils, fruit oils and butter played in the evolution of the human brain? In a word, none. Apart from the small amounts of fat that are present naturally in certain plant foods, for the majority of our time on this planet the only fat that humans consumed was animal fat.

To get an insight into how natural animal fats are in comparison with vegetable fats, you have to know something about how the cooking forms of these different fats are produced. Animal fats, such as dripping and lard, are melted down by heating them gently, and there is no need for preservatives or harsh chemicals. Butter is made simply by skimming cream off milk and churning it until it becomes solid. Clarified butter (also known as ghee) is made by gently heating butter and skimming off the milk solids. All are simple production processes which can be done in any kitchen. Fruit oils, such as olive, coconut and avocado oil, are made by mechanical extraction. This is not quite as simple, but it's still just about possible in your kitchen (although I don't know anyone who makes their own fruit oils). Seed oils, however, need more than a kitchen and a YouTube how-to video.

PRODUCTION OF SEED OILS

The final product may look wholesome …

... but its production needs an industrial plant.

TO PRODUCE ONE BOTTLE OF SEED OIL

Requirements: one industrial processing plant, seeds, bleach, hexane, phosphate, deodorizer.

Step 1	Crush seeds.
Step 2	Heat seeds to a very high temperature, typically 500°C; this oxidizes the oil, making it greyish in colour with a rancid smell.
Step 3	Bleach the oil to disguise the greyness.
Step 4	Soak in hexane or other petroleum solvent to extract the oil.
Step 5	Add acid to the oil to remove wax solids.
Step 6	Treat oil with chemicals to further improve the colour.
Step 7	Deodorize by steaming at very high temperature to remove foul-smelling residues.

Not possible in your kitchen.

TO PRODUCE ONE JAR OF HOME-MADE DRIPPING

Requirements: Beef fat, one knife, chopping board, hob and saucepan.

Step 1	Chop fat.
Step 2	Heat fat gently in a pan for a couple of hours.
Step 3	Pour liquid into jars.

Eminently possible in your kitchen.

We're being a bit unfair; these two examples aren't comparing like with like. The point we're trying to make is that animal fat can be prepared easily at home. Large-scale dripping or lard production for sale in a supermarket also requires an industrial plant, but such a plant carries out little more than the three steps above. Some brands add an antioxidant to extend the shelf life, but animal fats can be bought without additives.

In the nineteenth century animal fat comprised close to 100% of all fat consumed. By the 1980s vegetable oil comprised 80% of all fat consumption. How did this happen?

INDUSTRIAL VEGETABLE OILS: EATING COTTON AND THE CRISCO® STORY

It all started in the nineteenth century when two brothers-in-law, a Mr Procter and a Mr Gamble, had the bright idea of using the waste from cotton production to make cottonseed oil. First they used it to make oil for lamps, and then when electricity came along they had the bright idea of making cottonseed oil edible.

Thus was born Crisco®, the first commercially produced, edible cottonseed oil. Crisco was pearly white and solid and the name conjured up crispness and freshness. It looked just like lard, the most popular cooking fat of the time.

What isn't very well known is that cottonseed oil contains gossypol, a natural but highly toxic chemical. Gossypol causes infertility in men and has been used as a male contraceptive (in China); that is, until it was discovered that the infertility is irreversible. Gossypol is also a potent insecticide. Elimination of substances such as gossypol is one of the many reasons why seed oils need refining, bleaching and deodorizing before they are fit for human consumption. An issue of *Popular Science* at the turn of the twentieth century summed up the evolution of cottonseed: 'What was garbage in 1860 was fertilizer in 1870, cattle feed in 1880, and table food and many things else in 1890.'

In 1911, a clever marketing campaign ensured that Crisco became the predominant cooking oil in America over the next thirty years. The idea that Crisco was healthier than animal fats and that industrial processing was akin to purification was put about. This helped with the sales pitch that Crisco was a 'pure' all-vegetable product.

But the Crisco story was just beginning. In 1948, the American Heart Association, then not a very well-known organization, was the beneficiary of a Procter & Gamble-sponsored radio contest to the tune of $1.7 million (equivalent to around $20 million today). Thus the American Heart Association went from an underfunded society to a nationwide powerhouse. After receiving this money, the American Heart Association started promoting Crisco. Many are of the opinion that this funding of the American Heart Association and their promotion of Crisco as being heart healthy were directly connected. And so industrially produced 'vegetable oils' became mainstream.

Now we will get to the heart of why animal fats are healthier than vegetable oils.

FAT CONFUSION

Let's start with the difference between fats and oils. Chemically they are the same; but physically, at room temperature, fats are solid and oils are liquid.

Then let's consider why we need to eat fats. As far as the human body goes, fats are indispensable – they are the building blocks of certain hormones, including testosterone and oestrogen, and are needed for nutrient absorption, blood clotting, muscle movement, the proper functioning of the immune system, regulation of body temperature and the maintenance of healthy nails, hair and skin. The list goes on and on. Furthermore, fat is a very efficient way of storing and providing energy. Just as petrol or diesel fuels an internal combustion engine, so we burn fat to provide energy. And just as petrol is a more compact fuel than coal (it has a higher energy density), so too is fat more compact than glycogen (which is our short-term glucose store).

But doesn't eating fat make you fat? No! As outlined in chapter 5, eating fat on its own will not make you fat, even if you eat extremely large quantities of it (because eating fat by itself has virtually no effect on insulin levels). It's carbohydrates that raise insulin levels. So, no, fat doesn't make you fat, it's carbs that make you fat.

GOOD FAT, BAD FAT

If all the various types of fat have exactly the same effects on insulin, which they do, then why should it matter what type of fat you eat? As far as weight gain goes, no, it doesn't make a difference. But when it comes to general health, it does matter what type of fat you eat – there is a whole lot more to fat than purely its calories. This is because there is some fat in the outermost layer of every single cell in our body. The types of fat that we eat affect the composition of the fatty outer layer of cells (known as membranes), and hence affect the functioning and integrity of all cells.

This is especially true for our brain cells, which are covered in sheaths that are made of fat. These sheaths contain almost twice as much fat as the cell membranes in the rest of the body and they are crucial to keep the brain in working order. This is partly because the sheaths act as an insulating layer for the electrical signals that are transmitted along nerves (similar to the plastic coverings which insulate electrical wires). The fatty sheaths are what make the brain the fattiest organ in the human body (apart from our fat stores, that is). Even though our brains are capable of making most of the fats that they need out of glucose, the brain prefers to have readymade fats, particularly when it comes to certain omega fats (much more on these shortly). If these pre-assembled omega fats are already present in our diet, the brain can absorb them directly from the bloodstream rather than having to make them from scratch. So yes, to some extent, we are what we eat.

Now we will give the low-down on the different types of fat in our diet; they are all either saturated or unsaturated. Things start getting complicated when we subdivide these two fats. But rather than list them all, it is easier to use the diagram below. Whether you are eating butter, olive oil, soybean oil or beef fat, you are eating some combination of these various fats.

In the last few decades, we have been told that high levels of fat, especially saturated fat, in our diet increases blood cholesterol. And we have been told, incorrectly as it turns out, that high blood cholesterol levels are linked to heart

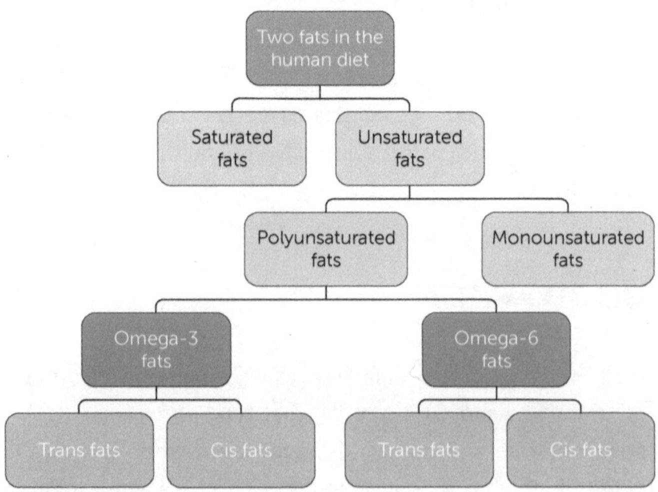

The different types of fat in our diet.

attacks. However, the USDA persist in recommending that we limit our saturated fat intake to 10% of calories. And where America goes, the rest of the world follows. Most European guidelines, including the UK, have similar recommendations. And because butter, dripping and red meat contain high levels of saturated fat, we have been advised to cut down on them and to replace them with vegetable oils, whole grains, fruits and vegetables which are low in saturated fat .

In a nutshell, we were told to swap saturated fat for vegetable oils and high-carbohydrate foods. But did you know that if your diet contains excess calories together with high levels of carbohydrates, these carbohydrates are converted into saturated fat? Or that if you eat a very low-fat diet, your body starts making saturated fat? What's the point of cutting down on saturated fats and eating more carbohydrates, if those carbohydrates will get turned into saturated fat anyway?

Next, let's take a look at cholesterol, that arch-villain of all the fats.

CHOLESTEROL, CHOLESTEROL TESTING AND THREE-LETTER INITIALISMS

Cholesterol – the fat that almost everyone has heard of. The very same one that supposedly causes heart attacks. Most think that the lower their cholesterol is, the better. What many might not know is exactly how essential cholesterol is in the human body – there is some cholesterol in every single cell of our body; it constitutes 30% of the protective outer layer of cells. It is so important that when we cut down on cholesterol-containing foods, our bodies make up for it by making more cholesterol.

Cholesterol is especially important for our brains. Cholesterol is abundant in the brain and is critical for brain plasticity; a lower cholesterol content in the brain is thought to impair cognitive function. And even though cholesterol in the blood can't get into the brain, the brain has its own independent cholesterol manufacturing system.

When a blood cholesterol test is done, the results include LDL (low-density lipoprotein cholesterol), HDL (high-density lipoprotein cholesterol) and triglyceride levels. LDL and HDL are actually something known as 'lipoproteins' which are a combination of fat and protein (this is because fat in blood has to be 'carried' on a protein). You have probably heard of LDL and HDL. LDL is sometimes known as 'bad cholesterol', and HDL as 'good cholesterol', although implying that there is such a thing as 'bad cholesterol' detracts from cholesterol's vital importance in the body. A high LDL often puts you on the conveyor belt for lifelong statin treatment (as almost happened to Mrs Taylor in the first chapter). True, a high-fat diet might push your LDL level up. But what many don't know is that there are many different types of LDL, and a high-fat diet pushes up the type of LDL that is not harmful, the large fluffy type of LDL (we're not making this up). More importantly, a high-fat diet, especially a high saturated-fat diet, lowers the worst type of LDL, known as small dense LDL. Out of all the different types of LDL, it is the small dense LDL that is most closely associated with heart disease.

Now we will look at triglycerides, which are the very worst of all the fats when it comes to heart disease (triglycerides have been shown to actually cause heart disease). What makes triglycerides go up? As we learned in the 'Cereal killers' chapter, it is carbohydrates. Our bodies convert excess carbohydrates into triglycerides. Also, in contrast to a high animal-fat diet, a high-carbohydrate diet increases abnormal inflammation. In general, the lower the carbohydrate content in a diet, the better.

All the different effects that fats and carbohydrates have on LDL can be a bit confusing, so here it is spelled out simply:

- High saturated-fat diets decrease harmful, small dense LDL = good effect
- High-fat diets decrease triglyceride levels = good effect
- High-carbohydrate diets increase small dense LDL = bad effect
- High-carbohydrate diets decrease HDL = bad effect
- High-carbohydrate diets increase triglyceride levels = bad effect

THE ESSENTIALITY OF FATS

Now we will take a look at polyunsaturated fats. Alongside advice to lower our saturated fat consumption, health authorities in the UK, USA, and many other countries also suggest increasing our polyunsaturated fat intake, because these are claimed to be healthier than saturated fats. But what exactly are polyunsaturated fats? You may have heard of something called omega fats – well, all omega fats are types of polyunsaturated fat (from here onwards the term 'polyunsaturated fat' will be used interchangeably with 'omega fats'). The main omega fats are omega-3 and omega-6 fats – there are others, but we need not concern ourselves with them here.

The big question is, are these omega fats really healthier than saturated fats? Let's first look at the risk of cancer if one consumes high amounts of omega fats. In humans, the evidence is mixed; some of the evidence suggests that high

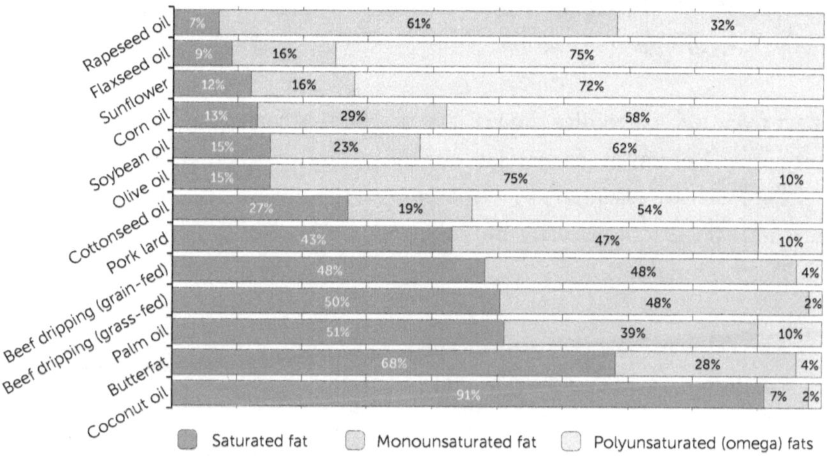

The ratios of the different types of fats.

levels of omega fats increase the risk of cancer, and some of the evidence suggests that they lower the risk. However, it has been known since the 1960s that excessive amounts of omega-6 fats cause cancer in animals. Corn oil, which contains more omega-6 than any other oil, has been routinely used in experiments to induce cancer in rats. The next question is whether these studies on animals are applicable to humans.

Let's suppose omega-6 were a new drug that was being tested by a drug company. They might start by testing out omega-6 on animals. If high levels of omega-6 fats were subsequently found to cause cancer in animals, it is unlikely that the drug company would get approval by any regulatory body to start testing it on humans. Even if drug companies were to get past the regulatory bodies (I wouldn't put anything past them), would you want to be taking a drug that is known to cause cancer in rats? Well, most people do consume large amounts of omega-6. Vegetable oils, including olive oil, contain ten to fifty times more omega-6 fats than do animal fats. Corn oil, soybean oil and sunflower oil have the highest levels of omega-6 fats of all. A small amount of omega-6 fat is an essential part of our diets. But massive amounts, as found in vegetable oils, are a problem. We would steer well clear of them.

The bar chart above shows the ratios of the different types of fats in commonly consumed fats.

To continue with the story of why replacing saturated fats with omega fats in our diet isn't a good idea, we need to explain a few other things.

OMEGA FATS AND YET MORE THREE-LETTER INITIALISMS

Apart from two particular omega fats, our bodies can make all the other fats that we need out of other foods in our diet. The two fats that our bodies can't make are termed 'essential', so we have to get them from our diet. Their names don't roll off the tongue easily, but here goes.

Two essential fats:

- ALA (alpha-linolenic acid), an omega-3 fat
- LA (linoleic acid), an omega-6 fat

We will now term these two fats (ALA and LA) the 'two essential fats'.

The two essential fats are important; but even more important are the other fats that our bodies make from them. Here we go again with a few more difficult words.

Fats made from the two essential fats:

- ALA is used to make EPA (eicosapentaenoic acid) and DHA (docosahexaenoic acid)
- LA is used to make ARA (arachidonic acid)

We will now term these three fats (EPA, DHA and ARA) the 'three very important fats', or 'VIP fats' for short.

More so than the two essential fats, it is actually the VIP fats that our bodies are constantly looking for and using. While the VIP fats can be made from the two essential fats, they can also be obtained readymade directly from certain foods. In fact, if we are consuming adequate amounts of the VIP fats from our diet, then the two essential fats may not be so essential after all. Why does this matter? Because, as we learned in chapter 2, we know that it was when readymade DHA (one of the VIP fats) entered our diets all those millions of years ago that our brains and intelligence started growing. We can go so far as to say that without sufficient DHA in our diet, human consciousness would never have been possible.

We will now cover a bit of basic biochemistry on DHA (one of the VIP fats) to show why what we eat matters so much for our brains. As we've said, we can get hold of DHA either by making it from ALA or by getting it from our food. Of significance here is that DHA is not found in *any* vegetable oil or in *any* plant food. Some algae (such as certain types of seaweed) and fungi (such as certain types of mushrooms) contain DHA, but the best sources are animal fats. And technically speaking, algae and fungi are neither plants nor animals, so it is absolutely correct to say that DHA is not found in any vegetable oil or in any plant food. (Just as it was correct to say that vitamin B12 is not found in any plant food.)

Because it is possible for our bodies to make VIP fats from essential fats (which can be found in plant foods) some say that it doesn't matter if we eat a purely plant-based diet. Trouble is, our bodies are extremely inefficient at this conversion (normally only 0.5% of ALA is converted to DHA). This pitiful rate of conversion is known to not provide enough DHA for the brains of babies and infants.

The diagram below is a visual representation of all these processes.

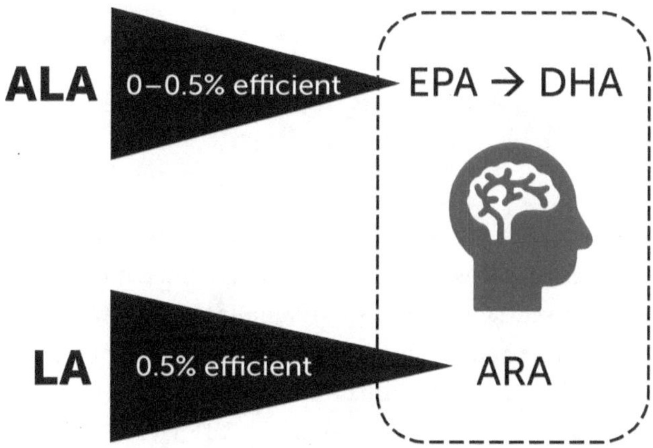

Supply of fats to the brain. **ALA and LA:** Found in animal and plant foods, **EPA, DHA and ARA:** Found only in animal foods, in certain seaweeds and in certain mushrooms.

The other big problem is that consuming large amounts of omega-6 fats can slow down or even shut down completely the conversion of ALA to DHA. As we now know, it is vegetable oils that contain massive amounts of omega-6 fats. Even olive oil isn't off the hook: it contains thirteen times as much omega-6 as beef fat. (Coconut oil is the rare exception – it is the only vegetable oil that doesn't contain high amounts of omega-6 fats, but coconut oil comes with other problems.) So forget buying trendy oils such as walnut or flaxseed oil in the hope that they will push up your DHA; even though they contain decent levels of ALA, the chances are that they won't make a blind bit of difference.

Luckily for us, animal fats contain readymade DHA – if we consume animal fats, we don't need to rely on the feeble or non-existent conversion of

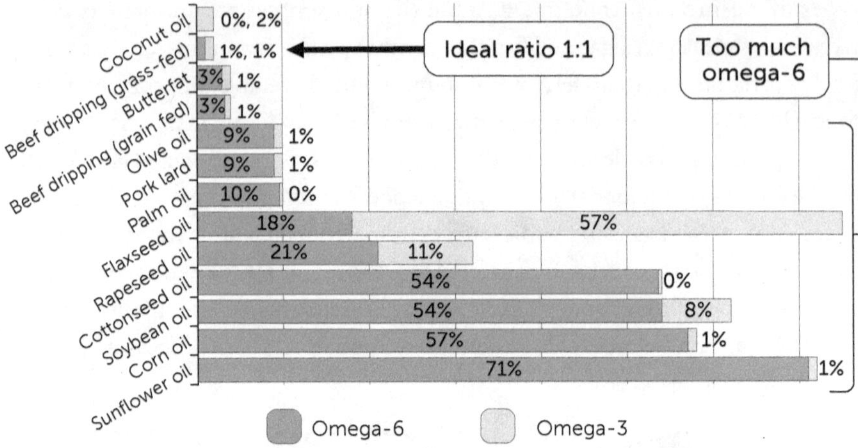

Percentage omega-3 and omega 6 in common fats and vegetable oils.

ALA to DHA. As far as the content of these fats in the human diet goes, for the majority of our time on this planet, we used to eat small but roughly equal amounts of the two essential fats plus small amounts of the readymade VIP fats. It was these amounts and combinations of fats in our diet that was partly responsible for the growth of our brains over the aeons. In fact, one-third of all the fat in our brains consists of just two of the VIP fats (DHA and ARA). For our brains to function properly a constant and sufficient supply of these VIP fats are needed. Importantly, the VIP fat supply to the brain depends on how much of them we have eaten.

Above is another fat-profile chart, this time showing the percentages of omega-3 and omega-6 fats. The ideal amounts of these two fats, in the right 1:1 ratio, is found in pasture-fed beef dripping. In contrast, vegetable oils contain far too much omega-6, both in absolute and relative terms.

OLIVE OIL – DEBUNKING THE DARLING OF THE MEDITERRANEAN DIET

It is now worth taking a brief look at olive oil to debunk the idea that it is one of the healthiest fats. So why exactly is olive oil thought to be healthy? Firstly, epidemiological studies show that the more olive oil one consumes, the lower is one's risk of heart disease and certain cancers. But as we learned in chapter 3, the results from epidemiological studies are unreliable. Another benefit of olive oil (also based on epidemiological studies) is down to the high monounsaturated fat level in olive oil. But monounsaturated fat is also one of the main components of beef fat, and beef fat comes without harmful amounts of omega-6. Yet another of olive oil's benefits is thought to be down to the powerful antioxidants (phenolic compounds, as they are known) that are in olive oil. But as we learned in chapter 8, our bodies don't need to be flooded with antioxidants. So you can swap that bottle of extortionately priced extra virgin olive oil for a reasonably priced jar of pasture-fed beef dripping while saving your health and some money.

WHEN OMEGA FATS FALL APART

Besides the issues discussed above, there are yet more problems with consuming large amounts of omega fats – they are unstable and vulnerable to damage by air, heat and light, which causes them to oxidize and form free radicals. The rancid smell we detect in foods that have gone off is owing to this oxidization of fats ('rancidity' differs from 'rotten' as it refers specifically to the smell or taste of fats that have oxidized). Foods that are at highest risk of going rancid are those which contain high amounts of omega fats. These foods include vegetable oils, fish oils, whole-grain products, nuts, nut oils or anything made with these ingredients (meat is relatively low in omega fat content and so doesn't go rancid as quickly as these other foods). Why does rancidity matter? It matters because when fats get oxidized, they trigger inflammation. Just as rust causes damage when metal gets oxidized, excessive inflammation causes damage in our bodies when fats get oxidized.

Then there's the effect of heat on omega-6 fats. This produces toxic compounds, with snappy names such as malondialdehyde, 4-hydroxy-2-nonenal and acrolein. Yes, they are as bad as they sound; they increase your risk of heart disease, joint disease, cancer and birth defects. Excessive consumption of deep-fried foods is regarded as a risk factor for diseases such as obesity, Alzheimer's, heart disease, cancer and diabetes. Most assume that it is the extra fat in fried foods that is the issue, but what really matters is the *type* of fat that foods are fried in. It is likely that it is these toxic compounds generated by deep frying that are harmful rather than the amount of fat.

Furthermore, it isn't just eating these toxic compounds that cause issues – even *inhaling* them, as occurs in those working near deep-fat fryers, causes problems. So if you do happen to find yourself overcome by the desire to deep-fry a Mars bar at any point, it would be least-worst to do it in beef dripping rather than in sunflower oil.

Now let's briefly come back to omega-3 fats. One of the supposed benefits of omega-3 fats is that they are 'anti-inflammatory' and a whole industry of omega-3 supplements has sprung up in the wake of this idea. However, as with all nutrients, there are downsides to consuming excessive amounts of anything – a very high intake of omega-3 fats can actually cause inflammation. Plus high amounts of omega-3 cause impairment of the immune system, abnormal blood clotting and may even promote the growth of certain cancers.

The bottom line is that animal fats are the best types of fat – they don't get damaged by exposure to air and light and are therefore much less susceptible to getting oxidized or becoming rancid (they contain large amounts of saturated fats, which are much more stable than omega fats).

This is the most technically challenging chapter in this book, so let's do a quick recap.

- Saturated fats lower the truly villainous small dense LDLs.
- Carbohydrates increase the villainous small dense LDLs.
- Saturated fats are stable at high temperatures and in bright light.
- Oils high in omega-6 fats induce cancer in animals.
- Vegetable oils (except for coconut oil) are all high in omega-6 fats, which reduce the body's ability to produce DHA, one of the most crucial fats for our brains.
- Omega fats are unstable at high temperatures and in bright light.

WHICH FATS AND HOW MUCH OF THEM SHOULD WE BE EATING?

If human studies show mixed results and we don't know how applicable to humans the animal studies on vegetable oils and cancer are, how should we know which fats and how much of them we should be eating? Well, for 99.9% of our 300,000-year existence as *Homo sapiens* – and well before dietary studies began – we have been eating the fat of large herbivorous animals. This type of fat, which is similar to grass-fed beef fat, typically contains 50% saturated fat, 48% monounsaturated fat and 1% each of omega-6 and omega-3; we evolved eating large amounts of saturated fat and relatively low but equal amounts of omega fats. It doesn't make sense that we now need to eat these fats in the opposite way – to cut down on saturated fat and increase omega fats. We should trust evolution over recent dietary studies, especially as modern studies can't tell us for sure what is the *ideal* amount or ratio of omega fats that we should be consuming.

We have seen that consuming vegetable oils has meant that nowadays we eat a lot more omega-6 than we used to. But another thing that has changed is the ratio of omega-6 to omega-3 consumed; this has gone up roughly twentyfold. Some scientists say that it is the change in the ratio of omega-6 to omega-3 that is harmful, so we should eat more omega-3 to improve the ratio. But actually, both the change in the ratio and the absolute increase in omega-6 fat are harmful – overconsumption of both of them is associated with an increased risk of inflammation, cancer, mental health disease and other diseases. The best way to get the level and the ratio right is not to increase omega-3 intake, but to reduce omega-6 consumption by eating animal fat and cutting out vegetable oils. How can we know for sure? One could do well-conducted dietary experiments; however, these are expensive, and difficult to carry out.

But evolution has already done the experiment for us. Ten thousand years ago, when we were eating only animal fats, and no grains, dairy foods or vegetable oils, our health was at its best. We were taller, we had no signs of malnutrition, and no evidence of cancer, arthritis, osteoporosis, dental decay or other diseases in our bones. At that time we were eating small but equal amounts of omega-3 and omega-6 fats. A natural conclusion would be that the new foods in our diet, including vegetable oils, have caused a deterioration of our health.

Another way to find out what fats we should be eating is by looking at what effects different fats eaten during pregnancy and breastfeeding have on babies. After all, evolution has been perfecting the recipe for breast milk since time immemorial. Of relevance here, though, is that what a mother eats, her baby also eats – if a mother eats large amounts of omega-6 fats, her breast milk will also have large amounts of omega-6 fats. No surprises there. What is a surprise is that the higher the levels of omega-6 fats in a mother's diet, the more likely it is that her baby's brain development will be impaired. Also, the higher the ratio of omega-6 to omega-3 in a mother's diet, the more likely it is that the child will have neurodevelopmental problems. As far as babies go, the absolute increase in omega-6 fat matters, as well as the change in the ratio of omega-6 to omega-3: both have negative effects.

The fact that high amounts of omega-6 fats are undeniably detrimental for a baby's brain combined with the 'evolutionary experiment' suggests that we should all be consuming smaller amounts of omega-6 fats. In a nutshell, we need to stop consuming vegetable oils altogether.

DHA – THE SECRET TO HUMAN INTELLIGENCE

Let's now take a look at DHA (one of the VIP fats) in breast milk. Why breast milk? Because DHA is a key promoter of brain development in babies, so the levels in breast milk matter a lot. In fact, the DHA level in breast milk is one of the best predictors of cognitive performance in children, better even than GDP per capita or educational expenditure – quite a staggering finding. Even more staggering is that there can be a tenfold variation in the amount of DHA present in breast milk – the level depends completely on the mother's diet. As far as diet goes, women following a plant-based diet have the lowest levels of DHA in their breast milk and those who include animal foods in their diet have the highest. Some think that if a mother has no animal food in her diet, the solution is to give her DHA supplements. Unfortunately, this is not as simple as it sounds, because DHA supplements cause other problems – they reduce the level of one of the other VIP fats (ARA).

However, it isn't just a low DHA that matters. It turns out that a low DHA together with a high omega-6 level is the worst combination for a baby's brain development. This combination is even worse than just a low DHA. Women who don't include any animal foods in their diets are probably also consuming vegetable oils – the worst combination for their babies. DHA is one of many, many necessary nutrients in our diet. Rather worryingly, prior to the 1990s, it was the norm for baby formula milk to have absolutely no DHA and to also have a very high level of omega-6. Because there are likely to be many nutrients in breast milk that we don't yet know about, it will always be difficult to formulate the perfect baby formula milk. Breast is best – if the mother includes animal foods in her diet.

But it isn't just during infancy that DHA matters, it matters throughout our lives. Adequate DHA in our diet has been shown to improve learning ability and a deficiency of DHA is associated with many problems, including ADHD, cognitive decline during ageing and Alzheimer's disease.

How much DHA exactly should we be consuming? I would go with the amounts of DHA that has been in our diet for the majority of our existence as *Homo sapiens*, rather than what some researcher pottering about in a petri dish says. As we learned in chapter 2, offal of ruminant animals is one of the best sources of DHA. Oily fish such as salmon are an excellent source too, but as fishing causes more damage to the ecosystem than grazing cows or sheep on green pastures do, we prefer the latter source. As we will find out in section two, it is better for the environment to eat regeneratively farmed meat and offal than to continue destroying the marine ecosystem by overfishing. Plus, offal is a whole lot cheaper than salmon.

Another quick recap.

- DHA (an omega-3 fat) was pivotal in the evolution of the human brain over millions of years.
- DHA accounts for 20% of fat in the brain, for instance in brain cell membranes and nerve sheaths.
- The body gets DHA either directly from animal foods in our diet or indirectly by making it from ALA (an omega-3 fat found in animal and plant foods) in our diet.
- The enzymes needed to make DHA from ALA are the same ones that are used to make ARA (an omega-6 fat) from LA (a supposedly essential omega-6 fat).
- Too much LA in the body impairs the ability to make DHA – there aren't enough enzymes to go round.
- Animal fats contain readymade and adequate amounts of DHA and ALA.
- Vegetable oils contain massive amounts of LA (an omega-6 fat), wrecking our ability to make DHA.
- Animal fats have small but the right amounts of all the omega fats that we need.

DHA IN THE DEVELOPED AND DEVELOPING WORLD

As already mentioned, DHA is important for normal brain function at all ages, but is particularly important in babies and in the elderly. A low level in older people is associated with an increased risk of dementia and cognitive decline. As we learned in chapter 9, a child's cognitive potential is permanently compromised by inadequate intakes of key nutrients before the age of five. International organizations such as the World Health Organization (WHO) have come up with the term 'first 1,000 days', which covers the nine months from conception and the first two years after birth. According to WHO: 'Having good nutrition in the first 1,000 days is critical not only to children's survival, but also to their development to reach their potential and become healthy and productive adults. When children face under-nutrition during this period, there are a range of consequences, often irreversible, which include frequent childhood illnesses, stunted growth, delay in development, and even death.'

Or as others have put it, 'A child's first 1,000 days of life are the most delicate and important for laying the foundation for their healthy growth and development.'

Clearly, nutrition is one of the major influences on overall growth and especially on brain growth, and DHA is one of the key nutrients for the brain. In babies, a low level is linked to poor visual cognition, for example. Around 250 million children in low- and middle-income countries are at risk of not fulfilling their cognitive potential. This has adverse long-term effects; these children do poorly in school and subsequently have low incomes, and this perpetuates the cycle of poverty.

Deficiencies of micronutrients in infants is an urgent global problem. Because the key nutrients for brain growth are found mainly in animal foods, pregnant and breastfeeding women in low- and middle-income countries are encouraged to eat animal foods by WHO. Yet in the UK and most other developed countries we are told that a plant-food only diet is fine for pregnant and breastfeeding women. How can it be that a plant-food only diet doesn't compromise the cognitive development of a child in a developed country, when it is recognized to do exactly that to children in low- and middle-income countries?

WE DID AS WE WERE TOLD

Forty years ago, the Americans and most Western countries were advised by government agencies to reduce cholesterol and fat (especially animal fat) and increase carbohydrate and vegetable oil consumption instead. In many respects, the Americans did as they were told (see table below). Of the 70g of added fats that Americans now consume every day, 60g are vegetable oils and 10g are animal fat. It is indeed bizarre that an increase in heart attack risk or diabetes is blamed on meat and animal fats – that blame should lie squarely at the door of vegetable oils and carbohydrates.

	Recommendations (percentage of total calories)	**Actually eaten over last 15 years**
Carbohydrate	45% to 65%	50%
Fat	20% to 35%	33%
Protein	10% to 15%	15%
Cholesterol	300mgs/day	276mg/day

CONCLUSION

Historically, small and equal amounts of animal-fat derived omega-6 and omega-3 fats used to be consumed, which was enough to supply all the omega fats that we need and, equally importantly, in the correct ratio. Nowadays, twenty times as much omega-6 compared to omega-3 is consumed. Animal foods, in contrast to vegetable oils, have ideal amounts and ratios of omega-3 and omega-6 fats.

DHA, an omega-3 fat that is especially important for our brain, is needed in small quantities, less than a gram per day. But not a single vegetable oil or plant food contains any DHA at all. Animal foods, however, offer just the right amounts of readymade DHA. If we had to pinpoint just one food that is vital for our bodies and for the environment, it would be offal, yet offal is notable by its absence from official dietary guidelines.

As a child growing up in India, I was encouraged to eat brain and fish because doing so apparently makes us clever. As a teenager I would pooh-pooh these claims as old wives' tales. Eating brain makes you brainy? Nonsense. However, now that I am older and wiser, and now that I know that brain and fish have more DHA than any other food, I would eat my teenage words, so to speak.

Long before olive oil, rapeseed oil, sunflower oil, corn oil and coconut oil appeared, animal fats were the only fat in our diet. Eating animal fats grew our brains. However, if one persists in following a purely plant-based diet – which we hope you won't after reading this book – you should at least include DHA-containing algae (such as seaweed) in your diet.

Vegetable oils are a slow poison. Just like cigarettes. One drop won't kill you, just as one cigarette won't kill you. But a slow steady drip over many years might. In an effort to improve our health and that of the planet, we have been advised to cut back on meat and animal fat, even though they are the most nutritious foods known to humankind, tried and tested by millions of years of evolution. Instead, we have been urged, by nutritionists and the food industry, to eat more vegetable oils, which are industrially produced, ultra-processed and full of polyunsaturated (i.e. omega) fats. Together with excessive carbohydrates, vegetable oils have increased our risk of obesity, diabetes, cancer, dementia and brain atrophy.

As with all poisons, it is the dose that matters. When saturated fats make up 50% of the fats in our diets, this is not a harmful amount for humans. While we all need some omega fats in our diets, especially the VIP fats (DHA, EPA and ARA), we need them in small amounts. Large amounts of omega fats, especially omega-6 fats, are harmful to humans. Less is more.

We have been told to switch from natural saturated fats to highly unnatural vegetable oils. As well as being industrially manufactured, vegetable oils are unstable, potentially toxic and mess up the supply of the fats needed by the brain. The USDA suggest 'use sparingly'. We urge 'don't use at all'. We're not sorry to remove them from the pyramid.

Although this brings us to the end of the demolition, we're not quite ready to clear the debris and begin afresh. That's because as well as a new pyramid based on the Sapiens diet, there are other diets that are high in fat and low in carbs – ketogenic diets – that could be considered. Before the builders move in, we need to assess the various ketogenic alternatives first.

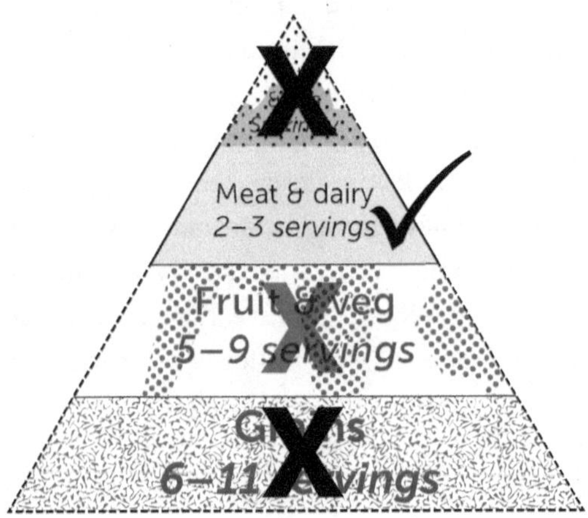

CHAPTER 11

THE KEY-TO-EVERYTHING DIET

The low-down on keto diets

'We need to embrace the idea that eating fat does not make us fat'
Nina Teicholz

THE KETOGENIC DIET IS A MODERN NUTRITIONIST'S nightmare – high in fat and low in carbohydrates. But keto diets are infinitely better than the diets that most nutritionists now recommend. For starters, high-fat 'keto' diets have been used successfully to treat epilepsy for over a hundred years. More recently, ketogenic diets have been found to have healing effects in many other conditions, from simple headaches to epilepsy, Parkinson's disease and diabetes.

To understand how a ketogenic diet exerts its benefits, we need to take a brief look at the body's energy systems. As far as the energy needs of cells go, all are able to use glucose as a fuel and most are able to switch to fat or another fuel when glucose is scarce. Ketones are one of the other fuels that most cells can use. And when one follows a ketogenic diet, your body produces ketones naturally. For this to happen, one's diet needs to be very high in fat, have very low amounts of carbohydrates (or none at all) and a moderate amount of protein. This kind of high-fat, low-carbohydrate diet, fuelling our bodies with ketones as well as glucose, is exactly what humans ate for hundreds of thousands of years during our evolution. So the ketogenic diet is no risky modern fad!

Trouble is, most people don't normally make ketones. Why? Because our bodies adapt to what we put in them. Cells use a particular fuel in proportion to its concentration in the bloodstream – if the concentration of a certain fuel goes up, cells will use that fuel in preference to others. Similarly, if the concentration of a fuel goes down, cells will use less of that fuel. If we are constantly eating large amounts of carbohydrates, our bodies will be continuously making large amounts of glucose. And having abundant amounts of glucose suppresses ketone production. Importantly, very high glucose levels in the blood are harmful so the body is forced to burn glucose to remove it. In this situation, which is the majority of time for most people, our bodies have no choice and use glucose as the predominant fuel.

The human brain, though, is a special case – there are a couple of issues with fuel supply to the brain. Most organs in the body can access and directly use fat from the bloodstream as a fuel. However, fat from the bloodstream can't get into the brain, so the brain can't directly use fat as a fuel like other organs can. Furthermore, the brain always needs *some* glucose. Trouble is, glucose reserves (in the form of glycogen) are fairly limited – roughly twelve hours of supply. So if you were to skip breakfast and your glucose reserve got used up while sleeping overnight (which it normally does), this could leave your brain high and dry. Luckily, biology has a few tricks up its sleeve.

Enter protein, which *can* be converted easily into glucose, a biological alchemy known as 'gluconeogenesis'. Fortunately, protein is always available in the form of muscle. However, using up large amounts of muscle is not a good idea; after all, during human evolution, muscles would have been vital both in running after the next meal and in running away from animals that wanted to make us their next meal. In an ideal world, we would have a fuel that reduces the brain's reliance on glucose, and so preserves muscle.

It so happens that there is such a fuel. Yes, indeed, it is ketones. These miraculous substances solve many problems for the brain. To start with they are made from fat, of which most people have more than adequate amounts. Furthermore, ketones can actually get into the brain from the bloodstream (unlike fat). What this means is that the brain has a way of using fat indirectly. As long as you have some fat, either in your diet or stored in your body, you will always have a supply of ketones. So your brain will be fuelled just fine if you are in a rush and don't have time for breakfast in the morning.

Hybrid cars are a good analogy to understand how the brain is powered, as both use more than one fuel. In hybrid cars, electricity and petrol are used, but the more electricity that is available, the less petrol is burned. Similarly, the brain uses glucose and ketones, but the more ketones that are available, the less glucose is used. If one is consuming a ketogenic diet, ketones become available and the brain uses both glucose and ketones. On a standard high-carbohydrate Western diet, ketones aren't available and the brain is forced to rely solely on glucose. It's the availability of ketones that determines how much glucose and ketones are used by the brain.

Let's now have a look at how the brain gets by with and without ketones. When ketones aren't available, the total amount of glucose that is needed by the brain is 120g per day. However, if significant amounts of ketones are available, the brain can get by with only 30g of glucose per day. As mentioned earlier, the magic of making glucose from non-carbohydrate foods kicks in when we aren't eating carbohydrates. Plus the much lower need of glucose on ketogenic diets translates into smaller amounts of it having to be made from non-carbohydrate sources. The bottom line is that the brain can always be kept well fuelled with a combination of glucose and ketones.

It is important to point out here that we don't need to rely solely on protein to supply glucose on ketogenic diets. When we are in ketosis and continuously breaking down fat, some of the breakdown products from fat also get converted into glucose. So after that overnight fast, there's no need to worry about massive amounts of muscle breaking down – fat and ketones kick in to save the day. All's well that ends well – so long as there is a fatty meal around the corner.

Apart from solving the brain's fuel problem during our evolution, ketones have recently turned out to be somewhat of a wonder drug. Some of the benefits of ketones are down to the fact that they stabilize one's brain chemistry. And besides ameliorating epilepsy, Parkinson's disease, obesity, autoimmune diseases, dementia and diabetes, ketones are also anti-inflammatory and act as antioxidants (so no need to search for kale or blueberries in the supermarket). Furthermore, ketones have also been used successfully in the treatment of certain cancers. But even if you don't suffer from any of these diseases, you can still benefit from ketones – they slow down ageing and have the life-extending properties of calorie-restriction diets – but without the hunger. Why suffer the hunger of a calorie-restricted diet if you can get their benefits while still feeling

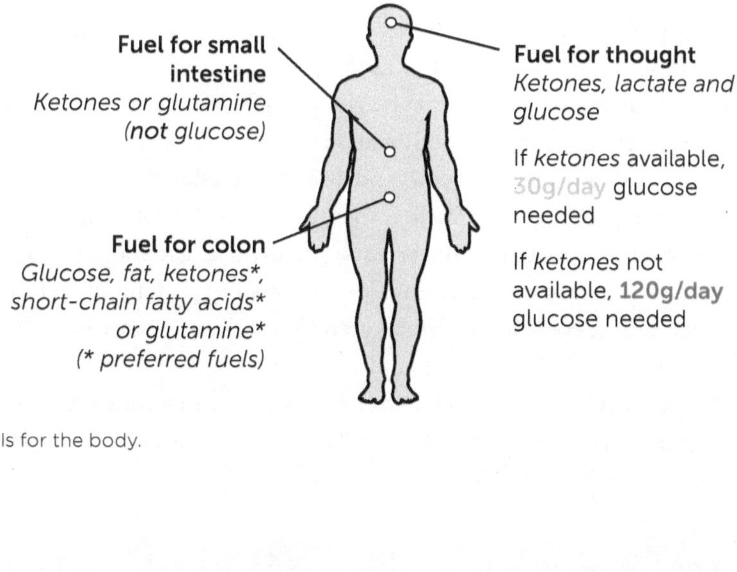

Fuels for the body.

full on a ketogenic diet? (My youngest daughter, the vegetarian one, responded to this rhetorical question with, 'Because bread is tasty ...')

Before we leave the subject of brain fuels, there is yet another important fuel, lactate (often called 'lactic acid') that the brain can use. Many think of lactate as being a waste product that causes muscle fatigue and soreness after exercising. But it turns out that the brain actively makes lactate even when we aren't exercising, and the brain *prefers* lactate over glucose when both are available. In fact, our bodies make and use lactate all the time and it can provide up to 25% of the brain's total energy demand. Some have dubbed lactate as the 'thinking molecule' because it is critical for laying down long-term memories, it alleviates depression, improves brain plasticity, protects brain cells and helps the brain heal after a stroke or traumatic brain injury. A long list of benefits. But there is yet more. Lactate is also a coordinator of whole-body metabolism and is vital in the development and growth of the baby brain.

Where do ketones fit in with lactate? Besides fuelling the brain and reducing its need for glucose, ketones and lactate can both act as 'signalling molecules' that relay messages between cells. Moreover, ketones can even replace the role of lactate in long-term memory formation. Ketones and lactate have both gone from once being considered a poison to now being viewed as a magic bullet.

WHICH COMES FIRST? CARBOHYDRATES OR PROTEIN OR FAT?

As we said earlier, the three macronutrients in our diet are carbohydrates, protein and fat, and these supply us with all our energy requirements. After we eat, the body first uses carbohydrates to make glucose; then, if there aren't enough carbohydrates in our diet to keep up with glucose requirements, the body switches to using protein to make glucose. If there isn't much protein in your diet, then as a last ditch the body starts converting fat to glucose (last ditch because it is an inefficient process and uses more energy than it makes). Importantly, ketones don't get made if there is too much protein in the diet – because excess protein can get converted to glucose. So, just as when too many carbohydrates suppress ketone production, so too does protein. In fact, this is why cats and other carnivorous animals aren't normally in ketosis; even though they don't eat any carbohydrates, their diets are too high in protein for them to routinely make ketones (compared to other animals, carnivorous animals have relatively large brains, which is why they have to eat large amounts of protein to ensure an adequate supply of glucose). A ketogenic diet for us humans, therefore, has to contain only moderate amounts of protein.

DON'T CONFUSE *KETOSIS* WITH *KETOACIDOSIS*

The state that the body is in when ketones are being made is known as 'ketosis'. When attained through a ketogenic diet it is termed 'physiological ketosis'.

However, ketosis has a bad association for many doctors, because their main association for the word is with a life-threatening condition known as 'diabetic ketoacidosis'. But diabetic ketoacidosis only occurs when there is insufficient insulin secretion, as in type 1 diabetes or in the very, very late stages of type 2 diabetes. This means ketone production goes unchecked, and excess ketones make the blood too acidic. It is this abnormal acidity that is life-threatening in diabetic ketoacidosis. So long as there is enough insulin, blood pH *never* goes outside the normal range on a ketogenic diet. In fact, a ketogenic diet can reverse type 2 diabetes (the most common type of diabetes) and can also help in controlling blood sugars in type 1 diabetes.

DOCTORS EXPERIMENTING ON THEMSELVES

Doctors who aren't well versed in the ketogenic diet think that it is unsafe for everyone, and particularly unsafe for type 1 diabetics. A couple of doctors who have type 1 diabetes decided to prove that this isn't the case by doing a fairly drastic experiment on themselves. What they did was walk twenty miles a day for five days; this was the doable part of their experiment. The challenge was to do the walk without consuming any food at all. For a full five days. In this experiment, what they found was that their ketone levels were similar to the levels in the non-diabetics who also did the experiment with them.

In a similarly extreme study, another type 1 diabetic undertook a 4,000km cycle ride across Australia over twenty days while on a ketogenic diet. This cyclist, too, arrived at his destination safe and sound, and kept excellent control of his diabetes during his ride. Even in these extreme conditions, physiological ketosis (as opposed to the dangerous type of ketosis) kicks in and is safe for type 1 diabetics. However, these intense physical challenges are not what I would recommend, whether you are diabetic or not.

It is true, though, that throughout the course of history, many famous, infamous (and ordinary!) doctors have conducted radical research, not only on themselves but also on their nearest and dearest. My children can testify to this. But to get back to the experiments of other doctors – some doctors experimented on their dogs in the days before insulin and they showed that dogs thrived on a meat-rich Inuit diet, whereas on a Hindu no-beef diet, the dogs soon went down with fatal diabetes. These results in dogs led to doctors trying out various diets on their diabetic relatives, and what they found was that an all-meat diet worked best in humans too. In fact, before the discovery of insulin, a very low-carbohydrate diet was the *only* treatment for diabetes. However, after the discovery of insulin, diets for diabetics have gone into reverse and high-carbohydrate diets have been recommended for many years now. Being on diabetic medication means that the effects of consumption of bread, chocolate and other such tasty high-carbohydrate foods may be mitigated, but in the long run these types of food cause a deterioration of diabetes.

IS THE KETOGENIC DIET 'NATURAL'?

A common question asked in relation to ketogenic diets is whether being in ketosis is 'natural' for humans. To answer this question, let's again look at newborn babies, whose brains account for an enormous 70% of their total energy needs (compared with 20% in adults). The baby brain is particularly vulnerable to a lack of energy, so much so that a shortfall rapidly causes brain damage. Nature, however, has found a way to mitigate this problem. Yes, ketones, once again. Breastfed babies routinely produce ketones, and if there is a temporary problem with the food supply, ketones step in to fill the gap. But mother nature has gone one step further with ketones – pregnant mothers start making ketones and sending them across the placenta just before their babies are born. From the moment a baby is born, it has a ready supply of a bespoke superfood. For human babies ketones are not only natural – they are necessary.

BEING FAT AND SMART

Humans, uniquely among terrestrial animals, have both an extremely large brain and a large amount of body fat. Compared to most other land-based animals, human babies have extraordinarily large fat stores – approximately 15% body fat compared with 2–3% in most other animals (with a few exceptions such as hibernating bears). It used to be thought that human babies evolved extra fat to compensate for a loss of body hair, i.e. to keep them warm.

However, some scientists have proposed that the modern human brain was a product of having first evolved fat babies and that the extra fat is there to provide the baby brain with energy (baby brains constitute 11% of their total body weight).

How exactly does a large supply of fat help a baby's brain? Firstly, it is these fat stores that make it possible for babies to produce ketones for their brains. Secondly, the baby brain has a phenomenal ability to use ketones in lieu of glucose – it is much more efficient at using ketones than the adult brain. Nature rarely does something without a reason – having large fat stores and being able to produce significant amounts of ketones are the human baby brain's insurance against starvation.

The fact that breastfed newborns are usually in ketosis has been one of the more surprising findings in my nutrition research over these past few years. It confirms that ketosis is a normal biological occurrence in babies. (I have been itching to test my grandson's blood for ketones ever since he was born. I will pluck up the courage to ask when my daughter's next child is born …) But is it also 'natural' for adult men and women to be in ketosis? If we look at ketosis in an evolutionary context, the evidence that ketosis is also natural in adults is compelling.

First, the ability to produce ketones – a fat-derived fuel – when glucose supply is low means that there is always a reserve fuel for the brain. Second, as humans began to evolve larger brains, further brain expansion would have depended on being able to increase ketone production. To guarantee a supply of ketones, fat stores would have had to increase as well. It turns out that it isn't only human babies that are fatter than the babies of other animals – adult humans are also fatter than the adults of most other species (even before the introduction of dietary guidelines!). Thus, having a good supply of fat and ketones is what would have enabled greater brain growth during human evolution. Other animals, however, weren't so lucky; their lower fat levels and inability to ramp up ketone production prevented them from evolving larger brains. If it weren't for our ability to produce and use ketones, even ten days without food would be fatal for most humans, rather than the sixty days on average that we can survive without food. Ketosis insured the human brain against starvation while also enabling the evolutionary brain expansion that is unique to humans.

Recent research backs up the evolutionary evidence, suggesting that the human brain *likes* using ketones. Studies show that, if ketones are available, the human brain will *always* use them. There is an astounding fivefold increase in ketone usage and a 33% reduction in glucose usage by the brain when our bodies are making ketones. This change in the ratio of fuel usage takes place even when there is adequate glucose. Moreover, ketones cause a 39% increase in blood flow to the brain. One can even think more clearly when following a ketogenic diet!

A HUMMER OR A TESLA?

Besides providing insurance for the human brain against starvation and enabling evolutionary brain expansion, ketones are also a 'cleaner' fuel than glucose. Let's use a car analogy again to explain this. The engine in our cells are called 'mitochondria' and this is where glucose and ketones are turned into energy. What exactly goes on in the mitochondria is extremely complicated and the subject of many large tomes of biochemistry. All that we need to know is that making energy from glucose produces lots of free radicals, whereas making energy from ketones actually reduces production of free radicals. As we saw in chapter 8, having too many free radicals is not a good thing. You could say making energy from glucose is like a slow and polluting Hummer and making energy from ketones is like a fast and clean Tesla.

WHO ACHIEVES KETOSIS MOST EASILY?

Besides not eating at all, there are other factors that affect how easily one goes into ketosis, such as the level of carbohydrate consumption, the length of time without food, age, how much exercise is taken and whether one is pregnant. Younger people and pregnant women go into ketosis most easily. Babies start making ketones if they have not had any milk for a couple of hours; children make ketones after several hours of not eating; young people do so after twelve to twenty-four hours of no food and it might take a longer period of fasting for older people. But after a ketogenic diet has been followed for several months, it gets easier and easier to make ketones, even in elderly people.

WHAT DOES A KETO DIET LOOK LIKE ON A PLATE?

As we have been saying, for a diet to be ketogenic we need to be eating a high-fat, moderate protein and very low-carbohydrate diet. Despite these limitations, ketogenic diets come in all sorts of shapes and sizes. In fact, it is possible to be pescatarian, vegetarian or even vegan on a ketogenic diet! Most people would be surprised to hear that you can be keto without eating any meat at all – although I wouldn't recommend any of the non-meat keto diets.

The reason why it is possible to be keto and meat-free is because fat, protein and carbs can come from many different foods. For example, the fat could come from vegetable oil (for vegans), butter (for vegetarians) or dripping (for meat eaters). And the protein could equally come from trout, a tofu burger or a truckle of cheese. Even more intriguingly, the few carbohydrates that one can consume on a keto diet could come from a sugary-sweet pudding (albeit a very small portion), a piece of bread or a bowl of strawberries. It has been a complete revelation that I could eat a chocolate brownie one day and still be in ketosis the next morning – although I wouldn't recommend eating brownies on a regular basis.

CONCLUSION

To sum up, a ketogenic diet is one that makes the body produce ketones. Typically, this happens when carbohydrate consumption is below a certain level; when there is no access to food for an extended period; or when one exercises to such a degree that all one's glucose stores get used up. Owing to the fact that ketones are made from fat, on ketogenic diets we need to replace the large amounts of carbohydrates that we currently eat with fat instead (and stick to moderate amounts of protein). A ketogenic diet shifts the body's metabolism into 'ketosis'; in this situation the body can indirectly use fat as a fuel.

There are many benefits to a ketogenic diet. Besides increasing longevity and ameliorating or curing many diseases, ketones are a 'cleaner' fuel than glucose – burning ketones doesn't produce as many free radicals as burning glucose does. Plus ketones decrease oxidative stress, increase antioxidants and scavenge free radicals. What matters to me, though, is that on a ketogenic diet one gets the favourable effects of fasting, without having to feel hungry.

We humans are unique in being able to use ketones regularly to fuel our brains. But our brain still needs a constant supply of glucose. On a ketogenic diet, the brain gets glucose by making it from protein, so there is no need to rely on carbohydrates. When ketones are available, they provide most of the brain's energy needs and the brain gets by with a much smaller amount of glucose.

The evidence from evolution, from babies, from recent good, controlled studies and from our own personal experience all point to a ketogenic diet being healthier and more natural than the standard recommended low-fat diets. However, even though most versions of a ketogenic diet are infinitely better than recommended diets, we should emphasize that many of them are not perfect – some versions may be deficient in vitamins and have been shown to cause stunting in growing children, and others may not be curative for autoimmune and other diseases (partly because the fat in these diets doesn't have to be animal fat and offal isn't compulsory).

But there is one version of a ketogenic diet – the Sapiens diet – that does contain the right amounts and types of fats, provides every single vitamin and mineral that the human body needs and has also been found to be curative for many diseases. More on the details of the Sapiens diet in chapter 14.

Next up, why gout – an excruciatingly painful affliction of the joints that is often put down to meat-rich diets – isn't an issue for the Sapiens diet.

CHAPTER 12

DOUBTS ABOUT GOUT

Why we don't need to worry about gout on the Sapiens diet

Doctor (to Hardy): That's the worst case of gout I ever saw.
Hardy: What causes that, doctor?
Doctor: Too much high living.
Laurel: Maybe we'd better move down to the basement.
Laurel and Hardy, *Them Thar Hills*, 1934

GOUT HAS DOGGED HUMANS SINCE ANTIQUITY AND has been called 'the disease of kings' owing to its tendency to afflict royalty and the wealthy. Most think of gout as an old-fashioned disease of the nobility that no one much suffers from anymore, but nothing could be further from the truth. Gout is, in fact, making a comeback these days.

The relevance of gout to this book is that the standard dietary advice currently given to gout sufferers is to cut down on their consumption of meat, offal, fish, alcohol and sugar, because these foods are thought to trigger gout attacks. This short, but still very important, chapter, specifically on gout, is to show that the Sapiens diet *does not* trigger or cause gout, even though it is a meat-based diet. Furthermore, the Sapiens diet has been shown to prevent further attacks in those who have previously suffered with the condition. To appreciate why this is the case, let's look at what causes gout.

Gout is caused by uric acid crystals and inflammation in joints. Uric acid crystals are produced when the uric acid level in the blood goes up. The key to preventing gout is to keep uric acid levels normal. The two things that influence uric acid levels are how much of it is being produced and how quickly it can be excreted. To properly understand this, we need to know how uric acid is produced.

Uric acid comes from the breakdown of a substance called 'purines'. Most of the purines in our bodies come from our own cells – as old dead cells are replaced with new healthy ones, purines get released. This is a continuous and ongoing process that we can't do anything about. As long as uric acid is excreted as quickly as it is made, there is no problem. At the end of the day, it's a balance between production and excretion, and there are many things that can upset this balance.

Much of the confusion about gout is because protein-rich foods make our bodies produce purines. When we look at the foods that make us produce purines (and hence uric acid), what matters the most is how they also affect uric acid excretion. So even though protein-rich foods contain purines and cause uric acid levels to go up, protein also speeds up uric acid excretion. As mentioned earlier, uric acid levels in the blood are determined by how quickly it is made and how quickly it is excreted; the net effect of eating meat, offal, fish and other high-protein foods is actually to *lower* uric acid levels in the blood.

The other thing to know about gout and uric acid is that insulin slows down uric acid excretion to a snail's pace. Even if you are not overweight or diabetic, any diet that increases insulin secretion will also increase the risk of gout. And because all carbohydrates cause insulin levels to go up, carbohydrate-rich foods will therefore also slow down uric acid excretion. This is why eating carbohydrates is more likely to cause gout than eating protein-rich foods. (And as mentioned in chapter 5, 80% of the population have very high insulin levels.)

The final culprits for gout are alcohol, fructose, all carbohydrate-rich foods and dehydration – they all slow down the excretion of uric acid and their net effect is to make uric acid levels go up.

How exactly does the Sapiens diet prevent gout? By lowering insulin levels, which speeds up uric acid excretion, resulting in lower uric acid levels in the blood. Plus, inflammation, which is one of the hallmarks of gout, is lowered on the Sapiens diet.

So, if you are prone to gout, go easy on the carbohydrates, alcohol and all foods rich in fructose (fruit, sugary foods and soda drinks). But don't worry about red meat, offal or fish – they do not cause gout when they are part of the Sapiens diet.

In sum, the biggest risk factors for gout are:

- High insulin levels – slows down uric acid excretion;
- Alcohol (especially beer) – slows down uric acid excretion;
- Fructose – slows down uric acid excretion and also increases insulin levels;
- Carbohydrates – increases insulin levels (which slows down uric acid excretion);
- Dehydration – slows down uric acid excretion.

Leaving aside what and how much one drinks, all these issues are avoided on the Sapiens diet. We now come to another of the key selling points for the diet. Why the Sapiens diet fixes a condition that many don't realize exists; a condition that we argue is the root cause of many diseases and conditions. Leaky gut.

CHAPTER 13

MIND THE GAP

The importance of intestinal permeability

'All disease begins in the gut.'
Hippocrates, 460BCE

MORE THAN LIKELY YOU'VE NEVER HEARD OF abnormal intestinal permeability (or leaky gut, as it is often known). If you search online for 'gut health' or 'intestine health', you'll be presented with lots of hits that talk about the importance of your 'gut microbiome' – things like having enough fibre in the diet, about feeding your gut biome with prebiotics and probiotics, and about 'loading up' on pulses, vegetables, fruit, and so on. But you have to search a bit deeper to find quotes like this:

> The majority of the cells that make up our immune system are found in our digestive tract. Having good gut health is linked to fewer sick days and lower risk of allergies and autoimmune conditions ... even if you put the healthiest food into your body, if you don't have a healthy intestinal lining to digest it, you won't get all the benefits of what you are eating.

This suggests that whatever the contents of the gut, and whether it's fed with prebiotics or probiotics, or loaded up with fibre, or with five-a-day of different-coloured fruit and veg, this counts for little unless the gut can do its job properly. And that requires a healthy gut wall. But it's interesting that even when the health and integrity of the gut wall is acknowledged, it's usually secondary, or almost an afterthought, to the ubiquitous 'gut microbiome'. It's hardly ever cited by itself as key to good digestion and good health. Indeed, when it comes to leaky gut, many health professionals deny such a condition exists. But there is a large body of science and research into leaky gut which shows that a healthy gut wall is one of the most important determinants of health. This chapter provides the latest evidence on leaky gut and health.

WHAT IS LEAKY GUT, AND IF IT IS A REAL CONDITION, WHY DOES IT MATTER?

It seems that every chapter has a bit of biology. I'm afraid this chapter is no different, so here goes. Inside our gut lies a barrier that protects us from the outside world. If it weren't for this barrier, anything passing through our gut, be it food or bacteria or anything else, would be able to enter our bodies. A healthy gut barrier acts as a security system which inspects every particle that goes past and only allows certain particles through. Ideally, nutrients should enter our bodies and all other unwanted substances should be kept out. You could say that when the gut barrier isn't working properly that it's as if the holes in the fly screen across the balcony door of your holiday apartment have become too big, and all sorts of unsavoury and annoying flies and bugs get in to bite your legs, feast on your tapas and drown in your Rioja.

So a healthy gut lining acts as a selective filter, allowing nutrients and water to enter the body while at the same time keeping unwanted substances out. This filtration function of the gut barrier is an important and often overlooked role of the gut. The malfunctioning of the gut barrier is commonly known as a 'leaky gut' (but technically known as 'abnormal intestinal permeability'). And the upshot of a leaky gut is that unwelcome substances are allowed to sneak into the body and certain wanted nutrients might get flushed away. All these events, in turn, lead to a decline in health.

Permeability is the term used to describe how easily things pass through or move within a substance. The sponge you use in the shower has high permeability – it's full of holes, and the holes are joined together, so water and soap can be easily absorbed and released. Bathroom tiles on the other hand have low permeability (or are 'impermeable'), keeping the water inside the shower cubicle, rather than letting it soak into the permeable plaster walls and then flooding the kitchen underneath. Likewise, *gut permeability* refers to how easily or not substances pass through the gut barrier. For ease of understanding we will use the term 'leaky gut' to refer to abnormally increased gut permeability.

The term 'leaky gut' needs some clarification. It does not mean that substances are leaking *into* the gut, but rather the other way around. When we say leaky gut, we mean substances from the hollow within the gut are passing into the body. Thus, leaky gut occurs when substances that are normally unable to cross the gut wall manage to do so and enter into the bloodstream.

STRUCTURE OF THE GUT WALL

The gut barrier is the largest interface in the human body. To give some idea how large that interface is, an average male adult has a skin surface area of around two square metres (2m²). If laid out flat, that would be enough to cover a typical dining table, though you probably wouldn't fancy sitting down at it for a meal. Whereas recent estimates for the gastrointestinal tract give it a surface area of 30–40m², or fifteen to twenty dining tables' worth, or about half a badminton court. (Older estimates had put the surface area as large as 200–300m², big enough for a game of doubles at tennis.) If we accept the conservative 30–40m², then the interface available to the gastrointestinal tract still dwarfs the area of the skin. By far the biggest contribution to the area comes from the small intestine (small so-called because of its relatively narrow diameter). The large intestine or colon contributes just two square metres, and the mouth, oesophagus and stomach only one square metre.

How is the small intestine area so large? Because the surface isn't flat and featureless. The gut wall is folded and wrinkly, which increases surface area for a start. Then, in addition, each wrinkle is covered in tiny protuberances, which in turn are covered in even smaller protuberances ('villi' and 'micro-villi'). Which all leads to a huge surface area coiled up inside our abdomen. And a large surface area makes the small intestine very good at its job of maximizing the absorption of nutrients from the food that we eat.

At the cellular level the surface of the small intestine is lined with a layer of mucus that forms a biochemical barrier – this is the first line of defence in the gut. Then a one-cell-thick layer behind the mucous acts as a physical barrier. And after that, immune cells lie behind the one-cell-thick layer and these form an immunological barrier; in fact, the gut immune cells form the largest immune organ in the human body. A hormone, zonulin, regulates the gaps between the cells that form the physical barrier. The mucous layer, the one-cell-thick layer, the regulatory hormones that control the openings between the cells, and the immune cells together constitute the gut barrier. Any substance trying to get through the gut and into the body has to traverse all these layers.

Substances that our bodies view as 'foreign' we call antigens. The immune cells in the gut barrier decide which substances passing through the gut are friendly or foreign. On the one hand, essential nutrients from food should be recognized as friendly and allowed into the body. On the other hand, harmful antigens should be recognized as foreign and kept out. When we do encounter harmful foreign substances, the immune system then makes antibodies to neutralize them. (What isn't well known is that certain foods contain substances that the human body sees as 'foreign'.) The gut immune cells are constantly performing a fine balancing act: between immune activation to protect against harmful substances, and immune tolerance to let in beneficial nutrients. A leaky gut is when this balancing act goes wrong.

WHAT CAUSES A LEAKY GUT?

There are many reasons why the intestine wall can become 'leaky'. One is if there is an abnormally increased level of zonulin, a hormone that controls the gaps between the gut cells. What always makes zonulin go up is gluten, something that is found in wheat and which most people eat every day in the form of bread. Furthermore, a lack of certain micronutrients, such as zinc, vitamin A and vitamin D, can also cause a leaky gut. Other causes include stress, allergic reactions, extreme exercise, gastrointestinal infections (especially in children in the developing world), high alcohol intake and medications like ibuprofen and some chemotherapy drugs. Even having abnormal insulin levels (without being overweight or having full-blown diabetes) is associated with leaky gut. The list goes on. Serious gut diseases such as Crohn's disease and ulcerative colitis give rise to a leaky gut, as can coeliac disease, which is triggered by gluten. The underlying biochemical mechanisms behind these causes are well established.

Of relevance here is that specific foods over-activate the release of zonulin, increasing the size of the gap between cells and causing leaky gut (especially gluten-containing foods, even in those who don't suffer from coeliac disease). But what isn't well known is that certain other constituents of food, such as wheat germ agglutinin, are viewed as foreign antigens by the gut immune cells. These other food constituents therefore also compromise the gut lining, allowing yet more foreign antigens to get in even more easily. The highest concentrations of wheat germ agglutinin are found in unprocessed wheat germ – which is present in foods such as muesli. Wholegrain cereals, muesli and legumes – the very foods that we have been urged to eat more of – are the very same foods that can cause a leaky gut. It's a wonder that we don't all have a leaky gut!

There are yet other foods that most of us eat regularly that are also linked to a leaky gut; foods rich in fructose, certain spices, medium-chain fatty acids (the type of fat in coconut oil) and omega-6 fatty acids (which are present in high amounts in all vegetable oils), to name but a few. Alcohol can also be a culprit for leaky gut. The effect of smoking on the gut is unclear. What about prebiotics and probiotics? There is no clear evidence that these provide any benefit for leaky gut. As for muesli and other similar wares of health food stores? There is little evidence that they are of benefit, and plenty of evidence that they might cause harm.

A summary of the causes of a leaky gut:

- Abnormally high level of zonulin
- Gluten
- A lack of certain micronutrients, such as zinc, vitamin A and vitamin D
- Stress
- Allergic reactions
- Extreme exercise
- Gastrointestinal infections
- High alcohol intake
- Medications such as ibuprofen and some chemotherapy drugs
- Chronic gut inflammation, Crohn's disease, ulcerative colitis
- Coeliac disease

LEAKY GUT AND AUTOIMMUNE DISEASES

A major role of the immune system is to defend us against diseases and to help us heal after any damage has occurred. But sometimes things go awry and our immune system starts mistakenly attacking our own cells. When this happens, the resulting conditions are known as 'autoimmune diseases'. Autoimmune diseases matter – roughly 10% of the population in the UK and the developed world suffer from some type of autoimmune disease (the prevalence is a bit lower in developing countries, but is on the rise).

Where does leaky gut fit into the autoimmune disease scenario? As mentioned earlier, antigens are normally kept out of the body as they pass through the gut. However, if one has a leaky gut these antigens can sometimes get through and enter our bodies (these antigens can be from food, from toxins or from bacteria). Some of these foreign antigens share similarities with our own normal body cells. And every so often the immune system is unable to distinguish between the antigens and our cells. What happens next is that the immune system thinks it needs to attack the cells which are similar to the antigens. It then starts making 'autoantibodies' to attack our own normal healthy cells.

For example, in type 1 diabetes the immune system makes autoantibodies against the insulin-producing cells in the pancreas. Other autoimmune diseases include rheumatoid arthritis, psoriasis, multiple sclerosis, SLE (systemic lupus erythematosus), ulcerative colitis and Graves' disease (a disease of the thyroid gland).

A leaky gut is also associated with leakiness of other barriers in the body such as in the brain, lungs, skin, genital tract and blood vessels. After the gut barrier, the blood–brain barrier is the one that has been studied the most. This matters too – a leaky blood–brain barrier is implicated in many neurological conditions, from epilepsy to brain cancer, autism, ADHD, multiple sclerosis, schizophrenia, dementia, Parkinson's disease and even migraine.

REVERSING LEAKY GUT – USP OF THE SAPIENS DIET

So far we've discussed what causes leaky gut and the diseases associated with it. But when it comes to diet, what's the good news – how can we cure or prevent leaky gut? Well, what we know for sure is that maintaining healthy levels of vitamin A, vitamin D, zinc and certain other micronutrients is very important, as a lack of these can definitely cause a leaky gut. And some studies show that weight loss in overweight and obese people is associated with an improvement in leaky gut. Most importantly, how do we know that medical conditions can be treated by correcting a leaky gut? Well, there is research which shows that this is, indeed, the case.

A medical research group in Hungary have been using what they call a paleoketo diet (and we call the Sapiens diet) to successfully treat autoimmune diseases, certain cancers, diabetes, obesity and metabolic syndrome in over ten thousand patients over the last eleven years. This research group have found that successful treatment of all these diseases is dependent on the correction of a leaky gut, on normalizing insulin levels and on correcting nutrient deficiencies. Some of their patients had a leaky gut, some had abnormal insulin levels and some had both. But all their patients with autoimmune diseases had a leaky gut. If these patients with autoimmune diseases then stopped the diet, their leaky gut would relapse, as would their disease. This was the case even if their nutrient status and insulin levels remained normal – it was leaky gut that was the most critical factor in autoimmune diseases.

The diet that they have been using for patients with serious medical conditions consists of meat, animal fat and offal. And absolutely nothing else. As their patients get better, they start reintroducing plant foods one by one to identify precisely which are the offending ones. For example, they had one patient with a condition known as alopecia universalis (loss of *all* body hair – from the scalp to eyebrows to pubic hair) and which is known to be an autoimmune condition that is difficult to treat. When this particular patient started the diet, one month later some hair started growing back in a small patch on their scalp. A couple of weeks later this patient drank a tiny amount of oat milk, and unfortunately the regrown hair fell out again. It was only when they then went back to keeping strictly to the diet that eventually their hair grew back again.

What this research group concluded was that cutting out the damaging factors that are present in plants and plant oils is what has a curative effect on leaky gut. To date, no other diet has been found to consistently rectify a leaky gut and also treat serious diseases. This is the unique selling point of the diet that we are advocating – it naturally ensures the integrity of the gut wall, and in doing so protects the body from the various, sometimes crippling, diseases and conditions that result from leaky gut.

HOW DO YOU KNOW WHETHER YOU HAVE A LEAKY GUT?

You might now be wondering: how does one know if one has a leaky gut? There are, in fact, several tests available. One of the more reliable tests is the polyethylene glycol 400 (PEG 400) test. The test uses an inert, harmless mixture that contains molecules that are chemically the same but vary in size. The person undergoing the test has to drink a solution with these molecules and then collect their urine six hours later. Their urine is then analysed to see what percentage of the different-sized molecules have been excreted in the urine. If the gut is leaky, then a higher percentage of molecules will get through and this can be measured in the urine. The Hungarian clinic has done countless before and after tests on their patients. What they have found is that when their patients keep to the diet,

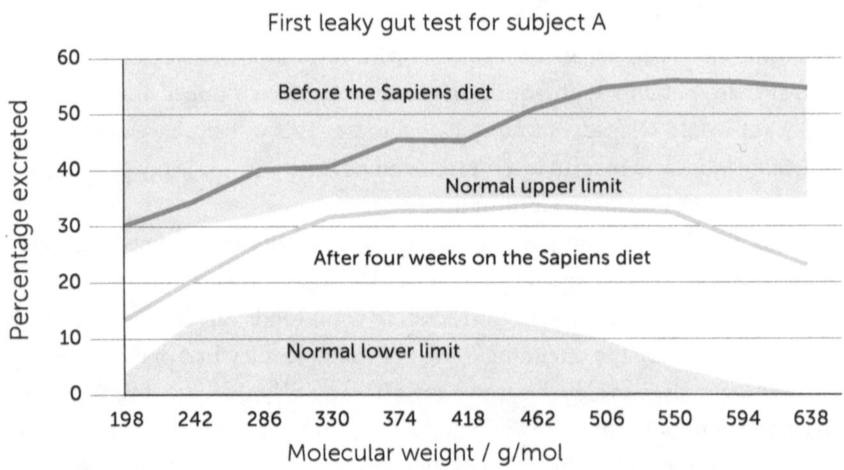

Leaky gut test before the Sapiens diet and after four weeks of the Sapiens diet.

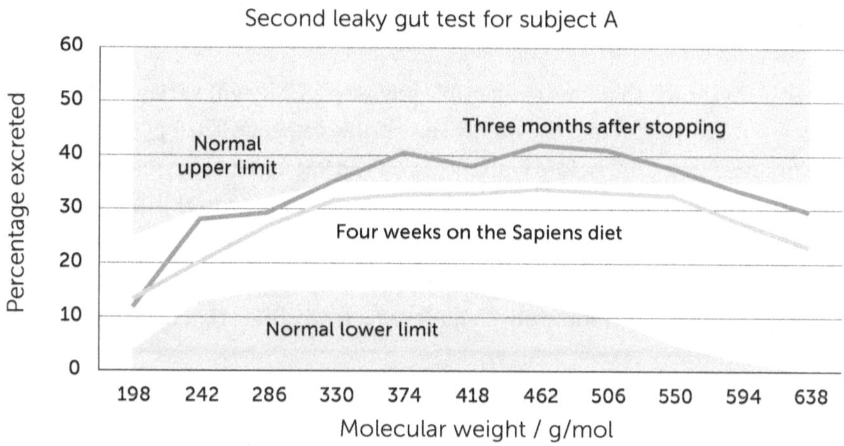

Leaky gut test after stopping the Sapiens diet.

their symptoms resolve and at the same time their leaky gut test becomes normal. But the improvements in their symptoms disappear if they stop the diet, and the leaky gut test becomes abnormal again.

The two figures (above and on p. 176) show the case where the leaky gut test was abnormal before starting the diet, became normal after following the diet for four weeks and then abnormal again three months after going back to a diet which included wholegrain cereals. The white area highlights the 'healthy' range.

LEAKY GUT AND THE MICROBIOME

Where does the microbiome come into all of this? There is a lack of awareness that the gut wall, and not just the gut contents, is important in itself. What would be the point of having a supposedly 'healthy microbiome' if all our barriers are leaking left, right and centre and letting substances that harm us into our bodies? What matters much more than the critters in our gut is the integrity of the gut wall and the integrity of all the other barriers in our bodies. And, as mentioned in chapter 6, most of the evidence on the benefits of the microbiome are poor-quality studies. Furthermore, butyrate (the darling of the short-chain fatty acid proponents) has been shown to actually worsen leaky gut.

Personally, I would prefer to have a cast iron, intact gut barrier with no holes in it rather than worry about the microbes in my gut – they can surely look after themselves without paying an exorbitant price for probiotics from a plastic bottle.

DOUBLE-CHECKING LEAKY GUT ON MYSELF

It has been shown that how susceptible one is to a leaky gut varies from person to person. Sadly, I fall into the highly susceptible category. The PEG 400 test was the first test that I did before embarking on the Sapiens diet myself seven years ago. To my surprise, I found that I had a leaky gut. I used to think I was eating very healthily; I had cut out grains, dairy and refined sugar for the previous two years and my diet consisted of large amounts of fruit and vegetables, a little bit of meat, fish and eggs, some olive oil (despite my dislike of it), nuts and copious amounts of tea.

After doing the initial leaky gut test, I cut out fruit, vegetables, nuts, olive oil and tea. I switched to a diet that was high in animal fat, had moderate amounts of meat, fish and eggs and small amounts of liver for four weeks. The next leaky gut test was normal. I was astonished to find that eating the exact opposite of official food guidelines caused a dramatic reversal of my leaky gut, especially as the standard thinking is that a high-fat diet causes a leaky gut.

However, standard thinking is based largely on studies on animals with diets which generally use vegetable oils or contain high amounts of sugar or carbohydrates or some combination of all of these. Studies which look specifically at vegetable oils alone find that they worsen leaky gut. More importantly, when high omega-6 fat vegetable oils are tested in humans, leaky gut is made worse, especially when the oil is heated to a high temperature, and including when the oil being tested is extra virgin olive oil.

CONCLUSION

Despite a leaky gut not currently being viewed as an abnormal health condition, *all* studies show that the integrity of the gut barrier is closely correlated with good health. There is strong evidence linking a leaky gut with obesity, type 1 and type 2 diabetes, cancer, autism, autoimmune diseases, eczema, irritable bowel syndrome and many other diseases. Some scientists believe that a leaky gut is the cause of these diseases, whereas others believe that a leaky gut is a consequence of disease. It may be that both views are partially correct.

Several studies show that a palaeoketo/Sapiens diet (which is high in animal fat) improves leaky gut. This is in contrast to many other studies which postulate that a high-fat diet is associated with a leaky gut. These conflicting findings may be due to the fact that vegetable oils and dairy foods can cause a leaky gut and these are often the source of the fats used in the studies that show a high-fat diet to be detrimental.

In essence, the gut wall is the body's bouncer and the integrity of the gut barrier is closely correlated with good health. Whichever way it is done, reversing a leaky gut is *always* associated with improved health. The Sapiens diet consistently repairs the gut barrier, and as a result reverses many diseases.

The leaky gut tests on myself, plus the evidence from the Hungarian research clinic, convinced me to write this book and to stay on the Sapiens diet as much as I can.

Adding the prevention of leaky gut to all the evidence of the benefits of the Sapiens diet from previous chapters means we are now in a position to finish this first section on diet and health by reconstructing the food pyramid.

CHAPTER 14

MEAT AND NO VEG

Why the Sapiens diet is the healthiest diet

'No disease that can be treated by diet should be treated with any other means.'

Maimonides

THE PREVIOUS CHAPTERS HAVE TAKEN A WRECKING ball to the original food pyramid. The grains and vegetable oils lie completely shattered, as does much of the fruit and vegetables layer, with just a small chunk preserved among the ruins. Only the meat and dairy layer remains intact.

This chapter will construct a new food pyramid, one that encapsulates the Sapiens diet. Before we begin rebuilding, we'll first explain why human health was at its best during the Palaeolithic era, and then we will summarize the key benefits of the Sapiens diet.

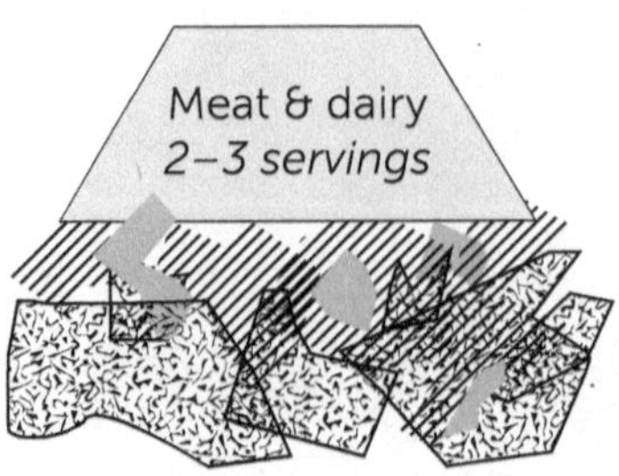

THE MAMMOTH IN THE ROOM – LIFE EXPECTANCY

It's all very well saying that when hunter-gatherers switched to agriculture their health deteriorated, but the big question is, who lived longer? The hunter-gatherers or the farmers? To answer this question properly, it is worth giving some background and going into a bit of detail of the lives of both of these groups of people.

Most people assume that hunter-gatherers were brutish uncultured people who died young and that the advent of agriculture and civilization meant an improvement in living standards, longevity and health. The reality, though, is complicated. Prior to the advent of agriculture some 10,000 years ago, small bands of nomadic hunter-gatherers (typically about 100 people) packed up and moved frequently. These itinerants were free to relocate to greener pastures and never had to worry about storing food – they were surrounded by food; all they had to do was hunt it down. Importantly, they left all their excrement behind every time they moved.

Farmers, on the other hand, settled down in one place, usually several hundreds of people in close proximity, and, to the delight of diarrhoeal diseases, their excrement also stayed in close proximity. Farms became a Pandora's box of pathogens and parasites. With the advent of farming, initially our health plunged, infection rates doubled, tuberculosis became common, intestinal parasites increased and dental decay arrived in force. Poorer nutrition also led to shorter people.

But even if farmers were short and ridden with disease, surely they must have lived longer than hunter-gatherers? Actually, in the first few millennia after the advent of farming, average life expectancy did not increase universally. For a select few, such as high priests and the ruling classes, life expectancy went up, but for most it went down.

To delve a bit deeper into this topic, let's now take a closer look at health and human lifespan across the world throughout history. Life expectancy for ancient Egyptians was twenty-three years, in classical Greece it was around thirty-five years, while in sixteenth-century England it was forty-two years. A serf working on their lord's estate and eking out a meagre existence in mediaeval times wouldn't have lived as long as the lady of the manor, who would have regularly dined on mutton served on a silver platter. There are extremes in modern times too. Currently life expectancy in Nigeria is a disappointing fifty-three years, whereas in Monaco it is an astounding eighty-seven years.

Who should we compare our Palaeolithic hunter-gatherers with? To a slave toiling at the pyramids 2,000 years ago, to a destitute person living in

Nigeria now or a millionaire from Monaco? If we were to compare the life expectancy of a typical Monégasque (probably driving a Maserati) with a hunter-gatherer (probably wielding a spear), the Monaco man or woman would be a hands-down winner. No surprises there. But it's a different story if we compare a hunter-gatherer to an ancient Egyptian labourer or someone in 2024 in war-torn Afghanistan.

WHAT ANCIENT BONES SHOW US

You might now be wondering how exactly do we know about the health and life expectancy of hunter-gatherers from several hundred thousand years ago when all we have are a few teeth and bones to look at? Well, there are scientists (palaeopathologists) out there who study these mummified remains in great detail and can identify exact diseases from their bones and teeth. The list of diseases that one can see in bones is fairly long. Tuberculosis, leprosy and syphilis are some of the infections that leave clear evidence in bones. Besides infections, there's rickets, scurvy, loss of height and porotic hyperostosis which can be seen in bones, and which are caused by deficiencies of vitamin D, vitamin C, vitamin B12 and other nutrients. In addition, one can also see arthritis, osteoporosis, fractures, bone tumours and the spread of cancer from other parts of the body in bones.

As far as teeth go, poor nutrition shows up as dental decay and 'stress lines' on teeth. And what the record shows is that, apart from traumatic fractures, all these ailments were non-existent in Palaeolithic hunter-gatherers. They arrived in force only after the advent of farming.

But out of the many markers of health in bones, height is one of the simplest indicators of general health. We see this in modern societies: when people from a less-developed country move to a more affluent one and their nutrition improves, the height of the next generation goes up. What ancient bones show is that there was an immediate and very clear drop in stature when our predecessors took up farming. This drop in stature persisted until recent times – it is only in the last century that average height has gone up again. Though average height has gone up and nutrition has improved considerably for many compared to the mediaeval era and the times of labourers building the pyramids, other markers of health have gone down substantially.

To come back to life expectancy. Yes, life expectancy matters. But health matters too. As a thought experiment, let's now try comparing the health of a hunter-gatherer woman from 100,000 years ago with that of a modern-day middle-aged woman from Monaco by looking at their bones. Their bones might both be equal in length, but they are unlikely to be equal in strength. The hunter-gatherer bones won't have osteoporosis, arthritis or a 'fragility fracture', but there is a good chance that the Monaco woman will have one of these (a fragility fracture is one that occurs when the bones have been weakened by an underlying condition, for example, osteoporosis). Calories can grow bones long, but you need other nutrients to keep them strong. Calories are available in abundance nowadays, but most of us don't go for the nutrient-rich foods that our ancestors routinely lived on. Overall, the mammoth hunters were tall and had no signs of malnutrition or degenerative diseases in their bones. There's no disputing their general good health.

The nub of the matter, though, is life expectancy. What palaeopathologists tell us is that, yes, on average our hunter-gatherer ancestors did die young. Most put the life expectancy of hunter-gatherers at thirty to forty years. Not fantastic. Even though the mammoth hunters were the best-nourished people in the entire history of mankind, clearly being well nourished isn't everything. Not many of us would be thrilled with the prospect of dying before the age of forty. Except, perhaps, rock stars.

Why exactly did hunter-gatherers die young? Mostly they died violent deaths; from accidents, from being pounced on by a puma or even by infanticide – 15–20% of all newborn babies in pre-agriculture times were killed to keep population numbers down. So, yes, many of them died very, very young. These brutal deaths, though, have skewed the average life expectancy of hunter-gatherers downwards considerably.

Then there is also the question of the accuracy of skeletal ageing techniques. Even though we can tell how well nourished 100,000-year-old hunter-gatherers were from their bones, it isn't quite so easy to see at what age they died. Some are of the opinion that the ages of older hunter-gatherers have been seriously underestimated.

We can also learn about longevity in hunter-gatherers by looking at extant populations (not many of these left) that eat in similar ways to hunter-gatherers from the Palaeolithic era. However, this is not a completely fair comparison because hunter-gatherers from recent times don't have free access to the land that they would have had 10,000 years ago. Plus modern-day environments have fewer large wild herbivores than there used to be. When questioned, hunter-gatherers from all parts of the world say that meat is at the top of their list of preferred foods, with honey a close second. Sadly, their current habitats are not teeming with the large fauna and wild beehives of yesteryear. And even if mammoths were still roaming the planet, it is likely that hunter-gatherers would be banned from hunting them down anyway.

But it isn't just what you eat that matters – what you don't eat also matters. Despite the limitations on their preferred foods, and not having their ideal diet, we do know what foods current hunter-gatherers don't eat. Bread, other cereals and vegetable oils are not on their menu. In some of the populations that have been studied in the last century, once these people make it past childhood, they live considerably longer than thirty years. There are even well-documented cases of some living to a hundred. When the health of the population of one of these cultures was studied in detail, there was not a single case of heart disease or stroke in the entire community. Nor any diabetes or dementia. And those who reached the age of sixty and beyond did so without suffering the chronic degenerative diseases that afflict the majority of the elderly in developed countries. So what do they die of? They die in their sleep of 'old age'. A good way to go.

At whatever age hunter-gatherers died, we know that they didn't suffer from malnutrition or other common ailments such as osteoporosis, arthritis or dental decay. And evidence of cancer in their bones was extremely rare. Moreover, malaria, which is one of the commonest causes of death in human history, only appeared after farmers started clearing forests (it was originally a treetop disease).

WHY DID WE TAKE UP FARMING?

Another big question that springs to mind is that if farming had such devastating consequences on our health, why did the mammoth hunters switch to farming? After all, farming was back-breaking work. Plus, there would always be the threat of deluge, drought, frost or fire. Furthermore, the foods from crops usually needed to be cooked in order to make them edible. Grains were particularly hard work – besides cooking, they needed to be threshed, ground and dried before putting them into safe storage places for protection from fungi and thieves. And why did the shift to farming occur almost simultaneously (though some dispute this) around the world?

According to most anthropologists, the mammoth hunters hunted the mammoths to extinction. When mammoths were available in abundance, the planet's carrying capacity was 15 million people. Then climate change came along as we emerged from the last glaciation; the weather warmed up (temperatures went up by a phenomenal 4°C), much of the tundra turned to forests, hunting became tricky and our preferred foods (large game) started decreasing dramatically with the change in climate. Mammoth steaks dropped off the menu. We had to start hunting smaller animals and fish, and foraging for more plant foods. Eventually, we resorted to farming. Fuelled by the newfound discovery of bread, our numbers started increasing substantially. And so we became 'bread heads'. As the anthropologist Mark Nathan Cohen says, 'People worldwide eat meat and various fruits when they can, and eat cereals and tubers only when they must.'

The bottom line is that one of the ways of increasing your life expectancy is with your diet. So why not eat in a way that maximizes your health?

PREVENTING AND TREATING MODERN DISEASES

To come back to the present day, we think it is better to treat the root cause of a disease than to take medication for the symptoms of that disease. Diabetes, high blood pressure and obesity, which are often the result of a Western-style diet, are commonly treated with medication, and the underlying causes are ignored. Medication usually masks the symptoms of these diseases and does not always cure the underlying disease. And obesity frequently gets treated with a diet that actually causes obesity.

The Sapiens diet helps in the commonest medical issues facing most of the world. It prevents and/or rectifies obesity, diabetes, pre-diabetes, cardiovascular disease, malnutrition, epilepsy and autoimmune diseases. It also has been used to successfully treat certain cancers and slow down or halt dementia in the early stages of the disease.

WHY WE NEED A NEW PYRAMID

Now we will summarize all the evidence that we've laid out in the previous chapters to explain why a new pyramid, based on the Sapiens diet, is desirable.

EVOLUTION

An animal-based diet drove our evolution since we split from the chimpanzee line some 6 million years ago, and resulted in the emergence of our species, *Homo sapiens*, 300,000 years ago. Meat, fat and offal are rich in the nutrients that helped trebled the size of our brains and grew our intelligence. Our digestive systems, with a relatively long small intestine and shortened colon, reflect our adaption from eating low-quality, plant-based foods to high-quality, animal-based foods. The archaeological record shows that when our species began to deviate from our ancestral diet some 10,000 years ago and we started consuming significant amounts of plant-based foods, our stature diminished, our bone and dental health deteriorated, and many modern diseases emerged. Since farming took root, average human brain size has shrunk by about 7%.

BAD SCIENCE

Poor-quality dietary studies have misinformed dietary advice for decades, from the US Senate Select Committee in the 1970s to the USDA food pyramid and now the current Eatwell plate. Many of the hypotheses from that flawed science have been disputed and disproved over and over again. Nevertheless, they persist; for example, dietary bodies, and the media, still assert that consumption of saturated fat causes heart attacks. The flawed conclusions from those studies are the unsound foundations of the original food pyramid, which senselessly endorses a diet that is the polar opposite of how we ate during our evolution.

CARBOHYDRATES

We now eat unprecedented quantities of carbohydrates and sugar compared to our pre-agriculture ancestors. Overconsumption of these is linked to diabetes, obesity, heart disease and dental decay. As if that isn't bad enough, gluten in grains, among other things, is linked to leaky gut, a condition which results in many autoimmune diseases. Fibre in plant foods has no nutritional value and carbohydrates are not an essential nutrient. Yet current nutritional advice advocates foods that are high in fibre and carbohydrates – grains and starchy vegetables – rather than the animal foods and fats that we are designed to consume. The Sapiens diet rejects carbohydrates as the cornerstone of diet; instead, it restores animal fat and meat to their rightful place in our diet. In addition to providing 'clean' energy, animal fats have the right mix of nutrients to support both mind and body.

FRUIT AND VEGETABLES

Guidelines like Eatwell recommend fruit and vegetables as 'a good source of vitamins and nutrients'. But the same guidelines admit that a plant-only diet contains absolutely no vitamins D or vitamin B12, and is deficient in selenium, iodine, calcium and iron. Other vitamins and nutrients in plants are often poorly absorbed by our digestive systems owing to the anti-nutrients plants contain. All in all, we can't rely on 'five-a-day'. Those on a plant-based diet can resort to artificial supplements or fortified foods. But a better and natural remedy is to consume animal foods. Offal provides superior nutrition compared to five-a-day – and all at a fraction of the price.

VEGETABLE OILS

Animal fat used to comprise almost all fat consumed, but now vegetable oils account for the majority of our fat intake. Vegetable oils played no part in our evolution – they are a modern invention, industrially manufactured and highly processed. They are nutritionally imbalanced and the high levels of omega-6 fats they contain are known to cause cancer in animals. The pyramid suggests 'sparingly'. We urge 'none at all'.

BUILDING THE SAPIENS DIET PYRAMID

PHASE 1 – THE BASE

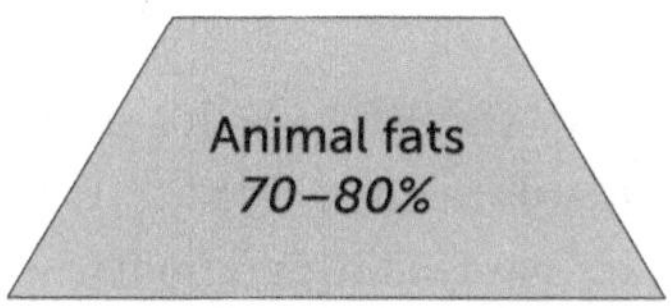

The body and brain need energy. Without it we can't move, or run our internal metabolic processes, or think. What we have learned is that it is better to be fuelling our bodies with ketones and glucose rather than with glucose alone. To be making ketones constantly, we need to be eating large amounts of fat. We have also learned that animal fats are superior and safer than vegetable oils. Accordingly, the Sapiens diet pyramid starts with a base of animal fat, which should comprise 70–80% of what we eat.

We imagine many readers may be dubious, seeing those percentages. But first remember – we do not get too fat by eating fat. We only do so when we eat carbohydrates instead of, or as well as, fat. And then remember – saturated fats in animal foods do not clog up arteries or cause heart disease. It's pancakes swimming in syrup, with or without blueberries scattered on top, that's a heart attack on a plate, *not* a fry-up of bacon and eggs. Just don't add toast to the fry-up.

And it's not to say that we should serve up huge chunks of dripping on the dinner plate, or pools of the same in a soup bowl. Yes, we can be generous with the dripping when frying or roasting, but much of the 70–80% of fat will be incorporated in the meat itself, as fatty streaks within the flesh of the meat, for instance. Sometimes the fat is obvious – in bacon rashers or along the sides of a steak or a lamb chop. But a lot of it will be invisible, such as in the yolk of eggs, or in the flesh of fish like salmon and mackerel.

Forget choosing lean cuts of meat all the time. Pick and cook meats with the fat attached, select streaky bacon and fatty sausages (try to avoid those with added preservatives or wheat), and go for full-fat mince rather than any low-fat nonsense. There are meal plans in the appendix to help you on your way.

PHASE 2: THE MEATY LAYER

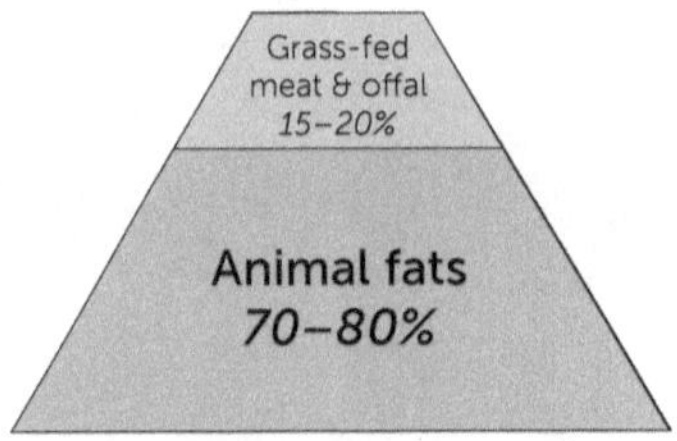

The next layer is meat and offal. We have this as its own layer, but as per the previous discussion, in terms of real foods – the things we actually eat – it overlaps to a large extent with the base of the pyramid; when we enjoy a fatty steak, pork belly or an untrimmed shoulder of lamb, we are getting both protein and fat, as well as a generous helping of essential nutrients. This layer also includes offal, which we have seen is a superb source of vitamins and micronutrients.

Together, these two layers provide:

- All the fat we need for energy.
- Fat with the right amounts of omega-3 and omega-6, including DHA, EPA and ARA, needed for a healthy brain and body.
- Protein to build muscle, and for the body to convert to glucose to help fuel the brain (along with ketones).
- All the vitamins that we need in bioavailable forms.
- Bioavailable minerals, including: iron, zinc, phosphorus, selenium, potassium, sodium, chloride, magnesium, iodine, sulphur, cobalt, copper, fluoride, calcium and manganese.

On a strict Sapiens diet, these are the only two layers we need. This strict form is recommended for people looking to put serious autoimmune diseases into remission without using drugs. In those with serious autoimmune diseases and allergies, eggs, fish, nuts, gluten-containing foods and many other plant foods can be problematic by causing a leaky gut; hence these foods are not included in the strict form of the pyramid. Only meat from purely grass-fed livestock is allowed in the strictest version of the diet (even meat from animals which have been fed soya can be a problem for some).

PHASE 3: OPTIONAL FRUIT AND VEGETABLES

Next we come to a less strict version of the diet which can still be used to treat obesity, insulin resistance, pre-diabetes and diabetes, though the strictest version will be quicker in ameliorating all these conditions. Now, for those who don't have allergies to fish and eggs, we can add them to the middle layer, and complete the pyramid with a pointy top.

The less strict Sapiens diet pyramid allows some fruit and vegetables, though not in significant quantities; more like one- or two-a-day than five-a-day. Eat the ones you like, though having said that, it's advisable to prefer

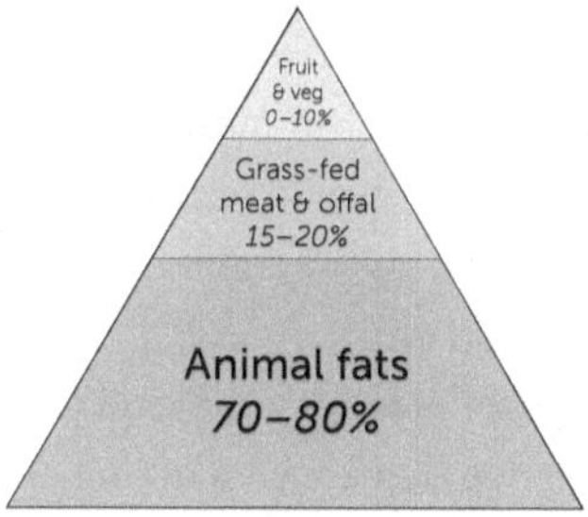

some and avoid others. Certain vegetables such as the nightshades are particularly undesirable because of the high levels and the variety of anti-nutrients that they contain (glycoalkaloids, capsaicinoids and lectins). The list of nightshades is long and includes tomatoes, aubergines, potatoes, peppers and goji berries (this is not an exhaustive list).

BEING REALISTIC

For those with absolutely no health problems we recommend keeping to the Sapiens diet as much as they can, ideally at least 80% of the time. Partially following this diet may not guarantee perfect health. But if the choice is between not following it at all or following it only some of the time, then the latter is a better option. And if we had to have a grading system of diets, we would recommend the following order:

1. Sapiens diet
2. Carnivore diet
3. Palaeolithic diet
4. Ketogenic diet
5. Low-carb diet
6. Mediterranean diet

7. Recommended guidelines diet
8. Standard American Diet
9. Plant-based diet

Details of the pros and cons of these different diets are summarized in the next chapter and meal plans for the top five diets are in the appendix.

BOLL*CKS TO BALANCE

We suspect some readers might still be sceptical. 'Yes, we've read and understand your arguments, that we eat too many carbs and too much sugar, that liver is nutritious, and so on. But surely cutting out all carbs and all or almost all fruit and vegetables can't be right. It means that the Sapiens diet *is not a balanced diet*.' Or words to that effect.

The phrase 'a balanced diet' is frequently bandied about. Often in the context of a story on a new way of eating that appears in a newspaper or online, and which concludes with something like 'Whatever the merits of this diet, we remind readers that it's best to aim to eat a balance of all the food groups.' For which read, 'We've no idea, so we'll play safe and offer no opinion whatsoever, and recommend "balance". That way we can't be completely wrong, and no one will sue us.'

We don't need a balanced diet. What we need is the *correct* diet. A balanced diet is a cop-out, a hedging of bets. A cow or a sheep doesn't need a balanced diet, it needs pasture, primarily grass (and, as we'll see later, no grains). A pet dog or cat doesn't need a varied diet of all the food groups; some of them would do it serious harm. It needs meat. So it is with humans. The Sapiens diet pyramid provides the right amounts of all the macro- and micro-nutrients we need, in bioavailable form, and without downsides such as insulin resistance or putting on weight.

To paraphrase a palaeo-dieter who tweeted recently on X: 'when someone tells me that I need a balanced diet with more carbohydrates, I respond: *what you're actually saying is that I need more insulin*'.

THE BEAR NECESSITIES

We are not alone in championing a move from the standard recommended diet to one that has almost no plant foods – there is now a growing band of carnivore diet followers. A carnivore diet is not dissimilar to the Sapiens diet – the main difference being that a carnivore diet can consist of only meat and doesn't have to include offal or high amounts of fat. Several high-profile people have made the change. One such is Bear Grylls.

Bear Grylls explained his switch from vegan to something close to the Sapiens diet in a recent interview with Louis Theroux. Theroux was incredulous that Grylls ate steak twice a day, with eggs and liver, and likewise astonished that Grylls questioned the 'five-a-day' mantra of orthodox nutritionists (on the basis that they contain anti-nutrients). Grylls was particularly dismissive of lentils: 'Terrible for you. I used to break wind all the time. I haven't broken wind in years now.'

We find two things interesting in this exchange. First, that Grylls had realized that his health 'had tanked' when he was vegan, and that switching to an animal-based diet was the answer. And secondly, that a seasoned broadcaster like Theroux seemed completely unaware, as are many in the media, that any doubts exist about the validity of current nutritional guidelines.

The carnivore diet is often successful in treating autoimmune diseases, diabetes and other serious diseases. However, there are some downsides to eating excessive amounts of protein and not including any offal in one's diet (such as deficiencies in certain vitamins). A few who haven't been successful on the carnivore diet have turned to something similar to the Sapiens diet and found that their symptoms then resolve.

CONCLUSION

The benefit of farming was the birth of civilization. But for the health of mankind, it was a disaster. We gravitated towards a class C diet without meaning to. One brimming with carbohydrates but barren in nutrients. The benefit of agriculture was its ability to support densely populated areas. In terms of population, agriculture has been a resounding success, supporting 8 billion people and counting (though some have too much, and others not enough, of the share). However, agriculture took us away from our optimum diet and optimum health. Moreover, industrial agriculture has an expiry date, one that's getting very close.

With the civilization hat on our heads, we have come to assume that pre-agriculture societies were unhealthy and malnourished. The reality is that civilized societies are the ones more likely to be unhealthy and malnourished.

Current guidelines recommend lowering fat, especially animal fat in our diet. The foods that have contributed to the calorie increase since the introduction of the guidelines are grains (9.5%), added fats – mostly vegetable fats (9%), added sugars (4.7%) and meat and nuts (1%). Of the 70g of added fats that

Americans now consume every day, 60g are vegetable oils and 10g are animal fat. Despite the fact that most of the fat increase is from vegetable oils, many think that the animal fat and meat part of the Standard American Diet has caused the current obesity and diabetes crisis.

The result of all this is that we have been the guinea pigs for an untested and experimental diet that ignores our evolutionary background. This worldwide experiment of a completely new set of nutrition recommendations was not registered and did not apply for ethics approval as all good research experiments are obliged to do. As we've said before, if the modern diet recommendations were a drug, they would never have been approved.

ONE SIZE FITS ALL

People may have had very different diets in the last 10,000 years, depending on which part of the globe one lives in and different cultural traditions, but for most of the previous 300,000 years of our existence, all *Homo sapiens* ate the same diet, and that is the diet that still suits us best. Not many have done well on the current recommended diets – the obesity and diabetes epidemic are testament to this. But *everyone* does well on the Sapiens diet – from Alaska to Zimbabwe, it is a safe and simple diet that suits every single one of us.

If someone whose health is severely compromised is about to start the Sapiens diet, then the strictest form of the diet is the best way forward, i.e. only animal fat, offal and meat. There are some people who are better adapted to plant foods than others and can therefore tolerate them better. However, most feel healthier when they cut out all plant foods; we would therefore advocate an all-meat diet, or as close to that as possible, for everyone.

Here's Eatwell versus the Sapiens diet, compared as plates.

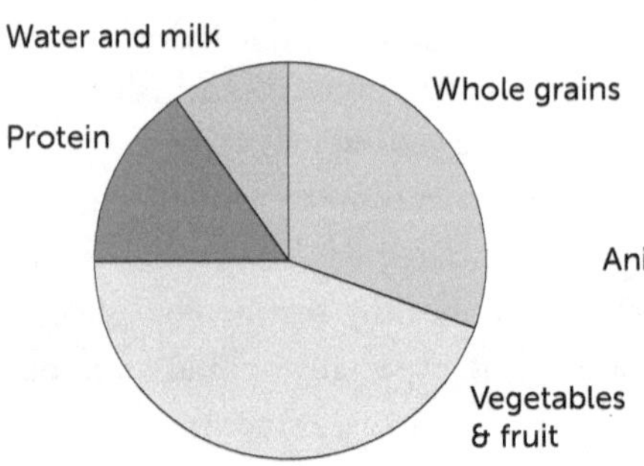

THE EATWELL PLATE

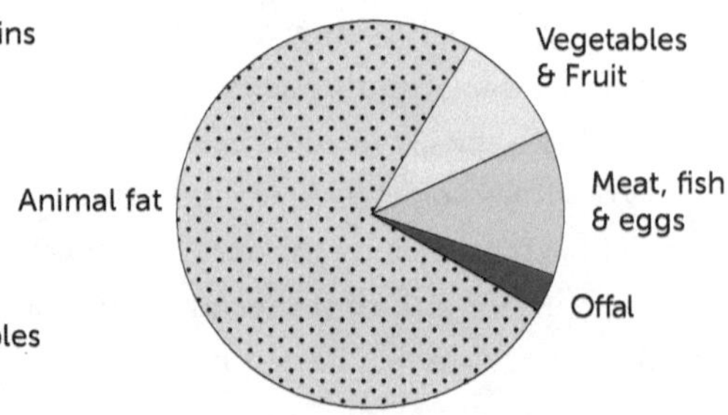

THE SAPIENS DIET

The four parts of the Sapiens diet plate consist of:

- Animal fat as occurs on the edges of steaks and chops and in bone marrow
- Meat, fish and eggs
- Offal
- Small and optional amounts of fruit and vegetables

Let's finally compare the original USDA pyramid and our Sapiens diet pyramid, side-by-side.

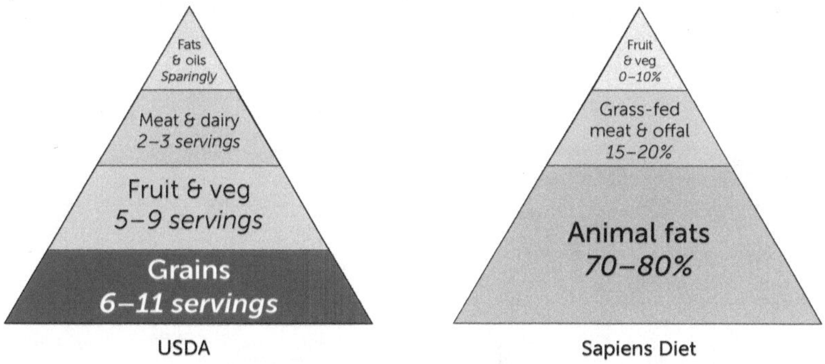

The new pyramid (right) is what Cartman from *South Park* would have wanted - an almost upside-down version of the original USDA food pyramid (left).

The three layers:

- The low-fat diet with vegetable oils at the top of the pyramid has become a high animal-fat diet at the base of the pyramid. Rather than 'fats and oils sparingly', we have 'animal fats liberally'.
- Instead of a carbohydrate-rich diet of grains, fruit and vegetables at the base, only small amounts of fruit and vegetables have now moved to the top of the pyramid as an option.
- The size of the meat layer in the middle stays the same, but now dairy has been removed and offal is included in it.

In sum, the Sapiens diet is the diet that we are best adapted to from an evolutionary point of view. It consists of animal fat, meat, small amounts of offal, fish, eggs, and small amounts of vegetables and fruit. All other foods are excluded, i.e. no grains, dairy products, legumes, refined sugar, vegetable oils, seed oils or nuts.

From grains and veg, to fat and offal.
From carbs growing our waistlines, to meat growing our brains.
We've come a long way.

But a bit further still to go. What we have shown in this section of the book is that, yes, we can stay healthy until we draw our very last breath. And that we also can carry on living in our sixties, seventies and beyond as we did in our twenties. And that the answer does lie on our breakfast, lunch and dinner plates. Another thing that is still left to consider is what different diets look like at mealtimes, which we will cover in the next chapter and in the appendix.

CHAPTER 15

DIETS DISSECTED

Summaries and pros and cons of some popular diets

'Finer meat ne'er met the eye, On the sweetest grass it fed.'
Jonathan Swift, *Mutton*

WE'LL NOW COMPARE THE SAPIENS DIET WITH the paleo, keto and carnivore diets.

These four diets have foods that are distinct to each diet, either being unique in type but also sometimes being unique in the quantity of the food that is recommended. It is also the case that each diet has foods in common with one or more of the other diets, such that they overlap each other. In the Venn diagrams that follow we indicate these areas in various shades of grey. The lightest grey indicates the area unique to that diet, the darker areas where there are overlaps. For these overlaps we highlight the types of food that the diets have in common. And for each diet we list the pros and cons in a table.

After comparing these diets, we give examples of typical weekly meal patterns and compare the carbohydrate contents. Then we summarise the key features of some other popular diets and conclude with a ranking of all the diets discussed, from best to worst.

THE SAPIENS DIET

Pros of the Sapiens diet	Cons of the Sapiens diet
• Healthiest of all the diets • Nutritionally complete • Makes you feel fuller than other diets • Reverses leaky gut • Puts type 2 diabetes, obesity and autoimmune diseases into remission • Reduces inflammation • Reduces chance of suffering physical or mental illnesses	• Can be difficult to keep to the diet in social situations • Likely to be met with bemusement or incomprehension in restaurants if you say that you don't want plant foods • While supermarkets promote a wide range of foods advertised as low-fat, they typically don't promote low-carb foods

Sapiens

- 15–20% protein (meat, fish, eggs)
- At least 70% of calories from animal fat
- Offal is compulsory
- Maximum of 10% of carbohydrates, in the form of fruit and vegetables
- No grains, dairy, legumes, seeds or nuts

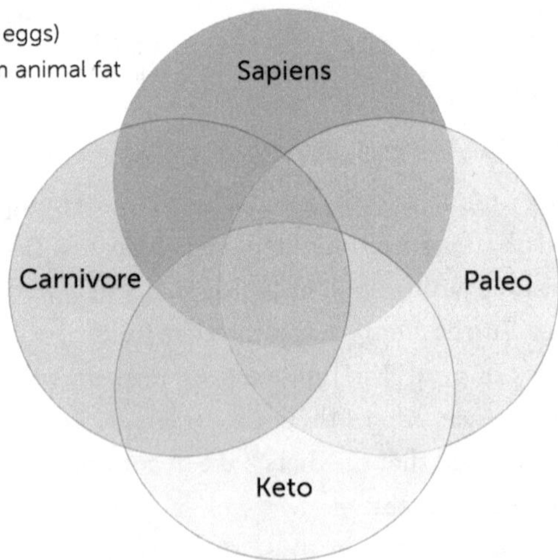

COMPARISON OF THE SAPIENS DIET WITH THE PALEO DIET

Pros of the paleo diet	Cons of the paleo diet
• Much healthier than government-recommended diets, Mediterranean diet, vegan and vegetarian diets, and Standard American Diet • Tastes good • Reduces risk of Western diseases • Makes you feel fuller than standard recommended diets, and SAD • Reduces excess body fat levels • High in vitamins and nutrients • Nutritionally complete if offal is included • Reduces risk for type 2 diabetes • Reduces inflammation	• You may miss eating non-paleo foods • Followers don't always eat 'nose-to-tail' as our ancestors would have done • Not necessarily nutritionally complete • Followers often include foods like olive oil or large amounts of seeds and nuts which, strictly speaking, isn't a Palaeolithic way of eating

Sapiens

- 15–20% protein (meat, fish, eggs)
- At least 70% of calories from animal fat
- Offal is compulsory
- Maximum of 10% of carbohydrates, in the form of fruit and vegetables
- No grains, dairy, legumes, seeds or nuts

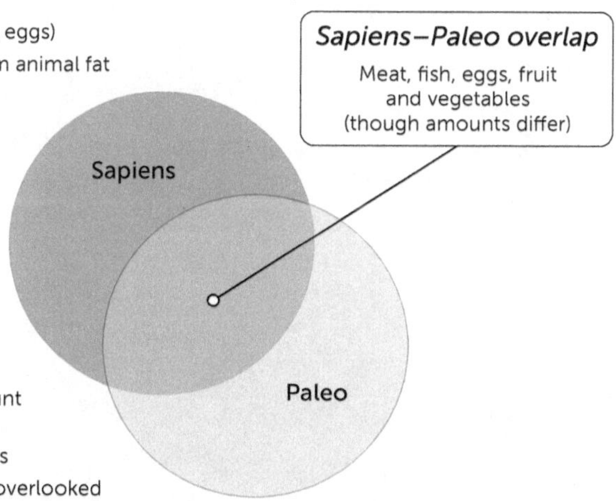

Paleo

- No specific limits on amount of protein
- No grains, dairy or legumes
- Importance of offal often overlooked
- Fruit, vegetables, seeds and nuts allowed

COMPARISON OF THE SAPIENS DIET WITH OTHER KETO DIETS

Pros of the keto diet	Cons of the keto diet
• Healthier than government-recommended diets, Mediterranean diet, vegan and vegetarian diets, and Standard American Diet • Well-established use in the treatment of drug-resistant epilepsy in children • Some evidence that the diet is effective at treating other neurological conditions • Lose weight without feeling hungry	• Associated with micronutrient deficiencies depending on food choices • Not always 100% effective at treating epilepsy

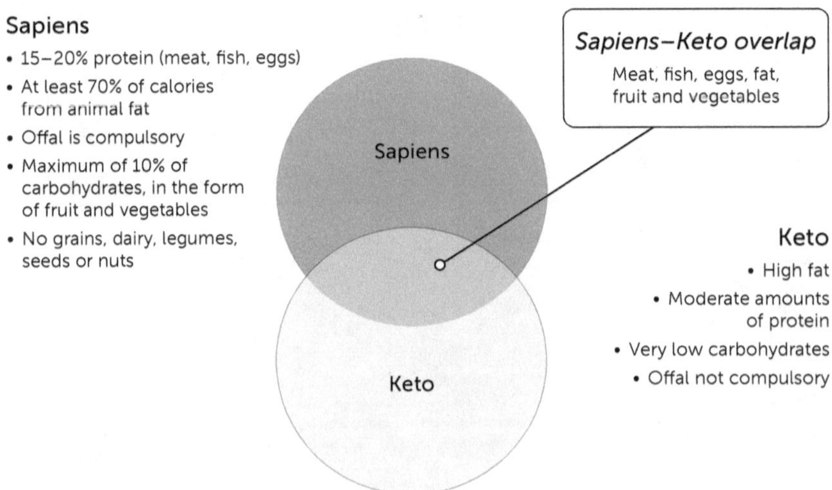

COMPARISON WITH CARNIVORE DIET

Pros of the carnivore diet	Cons of the carnivore diet
• Healthier than the Standard American Diet • Makes you feel good • Makes you feel fuller than government-recommended diets and Standard American Diet • Puts type 2 diabetes into remission • Often successful in putting autoimmune diseases into remission	• Possible to eat excessive amounts of protein • Isn't ketogenic if very large amounts of protein consumed • Associated with micronutrient deficiencies if offal isn't included

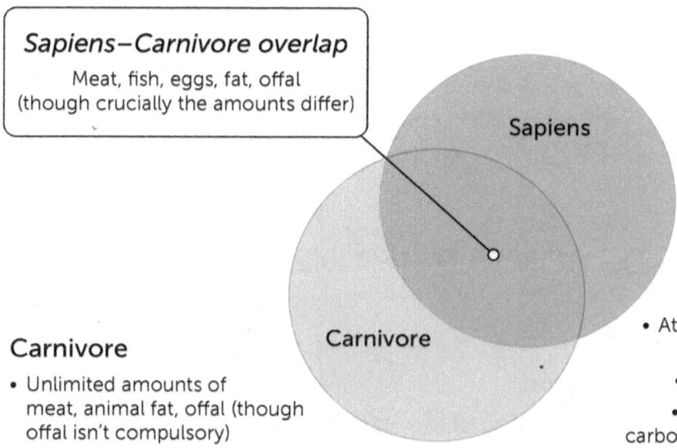

Carnivore

- Unlimited amounts of meat, animal fat, offal (though offal isn't compulsory)
- No specific fat-to-protein ratio
- No carbohydrates

Sapiens

- 15–20% protein (meat, fish, eggs)
- At least 70% of calories from animal fat
- Offal is compulsory
- Maximum of 10% of carbohydrates, in the form of fruit and vegetables
- No grains, dairy, legumes, seeds or nuts

TYPICAL WEEKLY MEAL PATTERNS

Sapiens	Fatty meat, liver, nitrate-free bacon, eggs, beef dripping or lard, bone broth, roast bone marrow, optional few low-carb vegetables, optional few low-carb fruits, optional one or two teaspoons of honey per day.
Keto	Fatty meat, fish, chicken, eggs, butter, vegetable oil, dripping, nuts, low-carb fruit and vegetables.
Paleo	Eggs, meat, fish, fruit, vegetables, nuts and small amounts of honey.
Carnivore	Steak, mince, roast joints.

CARBOHYDRATE AMOUNTS OF SAPIENS, KETO, CARNIVORE, LOW-CARB AND UK EATWELL DIETS

Carnivore diet	0g/day
Keto and Sapiens diets	0–50g/day (ideally less than 30g)
On low-carb diets*	0–100g/day
UK Eatwell diet	At least 333g/day for adult men At least 267g/day for adult women

* There is no strict definition of these diets.

SUMMARIES OF SOME OTHER DIETS

LOW-CARB DIETS

- Limits foods high in carbs.
- There is no standard definition of low-carb diets – there are many different types of low-carb diets.
- Carbs are usually replaced with protein or fat or both.
- It doesn't matter which foods provide the fat, protein or carbohydrates on a low-carb diet.
- Overlaps with keto diets if it is a very low-carb diet (less than 50g/day).
- Offal rarely featured.

Typical week: Meat, cheese, eggs, fish, olive oil, butter, low-carb vegetables, berries.

Pros of the low-carb diets	Cons of the low-carb diets
Healthier than government-recommended diets, the Mediterranean diet, the SAD and vegan and vegetarian diets	Associated with micronutrient deficiencies depending on food choices

US and UK Government-Recommended Diets, and the 'Mediterranean' Diet

- Low-fat and high-carb (from whole grains) diets.
- Saturated fat, trans fats, meat, salt, added sugars and alcohol are limited.
- Olive oil or other unsaturated oils are preferred over butter.
- Offal overlooked.

Typical week: Wholemeal bread, pasta, lentils, fish twice a week, small amount of lean meat, low-fat yogurt, salad with olive oil dressing, a handful of nuts, at least five portions a day of fruit and vegetables.

Pros of government-recommended and Mediterranean diets	Cons of government-recommended and Mediterranean diets
• Includes some meat, fish and eggs • Limits alcohol consumption • Discourages large amounts of sugar intake	• Encourages carb addiction • Introduction of low-fat guidelines coincides with the beginning of the obesity epidemic • Evidence on which the guidelines are based is highly questionable • Discourages consumption of healthy foods such as animal fat and red meat • No mention of offal – the most nutritious of all foods • Encourages consumption of unhealthy foods such as grains, legumes, dairy products and vegetable oils • Associated with micronutrient deficiencies (depending on food choices) • Encourages waste: large amounts of nutritious animal fat are routinely thrown away • Many U-turns in the last twenty years, e.g. on trans-fat laden margarines and on eggs • Difficult to trust the recommendations

Vegetarian Diet

A plant-based diet that excludes meat and fish.

Pros of a vegetarian diet	Cons of a vegetarian diet
If includes plenty of eggs and dairy products may be nutritionally complete and better than the Standard American Diet and vegan diets	If doesn't include many eggs or dairy products may be nutritionally incomplete and lack essential nutrients

VEGAN DIET WITH SUPPLEMENTS

A plant-based diet that excludes all animal foods, including eggs, dairy products and honey.

Pros of a vegan diet	Cons of a vegan diet
If low or moderate amounts of carbohydrates consumed, and if the diet is supplemented adequately, then it is healthier than the Standard American Diet	Lacks essential nutrients Catastrophic health impacts if not supplemented

STANDARD AMERICAN DIET (OFTEN KNOWN AS SAD)

- This is the high-carb diet that most people in Western countries actually eat.
- Characterized by excess calorie and carbohydrate consumption.
- Offal rarely featured.

Typical week: Burgers, pizza, pasta, chips, doughnuts, Coca-Cola and beer.

Pros of Standard American Diet	Cons of Sstandard American Diet
It tastes very good!	It is associated with an increased risk of obesity, diabetes, cancer and mortality Carbs are addictive

GRADING OF VARIOUS DIETS, FROM BEST TO WORST

1. Sapiens
2. Carnivore
3. Paleo
4. Keto
5. Low-carb diets
6. Mediterranean*
6. Recommended guidelines*
6. Vegetarian*
6. Vegan*
10. Standard American Diet

* Difficult to differentiate – depends on exactly what foods are eaten and in what amounts.

We have now reached the end of the nutrition part of the book. But what are the implications of recommending a high animal-sourced diet and reducing our intake of plant-based foods on farming? In the final section we will look at the impact of the Sapiens diet on farming and the environment. What we will learn is that livestock farming, to provide meat, fat and offal, is also highly beneficial to the health of the planet.

Section Three

How Cows Reverse Climate Change

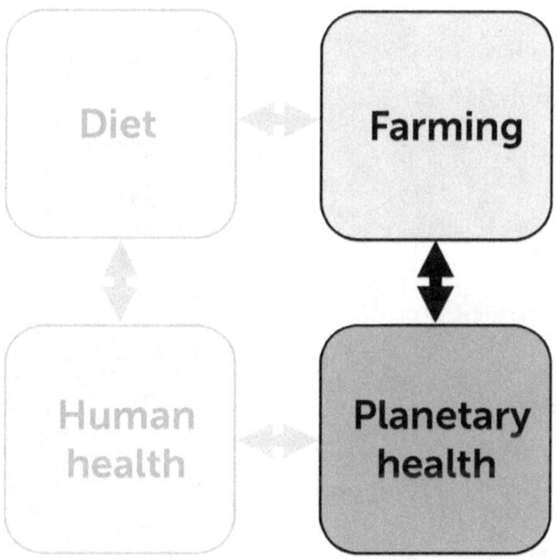

CHAPTER 16

FENS, MOORS AND MONGOLIA

Another author's personal story

'Pastoralists engage with livestock in ways that are simultaneously good for animals, people and planet. They are essential to upholding the web of life on Earth and ensuring its future functioning.'

Dr Fred Provenza, Professor Emeritus, Department of Wildland Resources, Utah State University

IT WAS A SULTRY AUGUST DAY. THE kitchen of my new home was filled with frantic droning again. I cringed at the sight of more dead flies littering the windowsill, some still spinning toes up towards eternity. Escaping into the garden, my nostrils were filled with the stench of rotting cabbage from the field behind. As far as my eyes could see was a flat, hedge-less, tree-less expanse of brassicas, wheat and oilseed crops. The house had been full of flies from the cabbages ever since I had moved to the Lincolnshire fenland village a few weeks before. Then, as the tractor rolled past, pulling a sprayer, I recognized a second smell – Metasystox. After just a few weeks in my first job as a crop agronomist, I was already familiar with the heavy odour of this insecticide.

'Right,' said my new colleague, Denis, 'can't go to that farm today, 'cause he's spraying Metasystox!' Denis had asked all his farmer clients to warn him when they were spraying, because Metasystox made him so ill. I marvelled that Denis had remained in the job at all and had become such an expert agronomist; it seemed that the whole county was busy with sprayers all summer long, trailing this pungent lethal mist behind them. And when they weren't spraying insecticides, every inch of crop and soil was being drenched with fungicides and herbicides.

Later that day I popped across to the plant health labs at our office, where I found the plant pathologist (a specialist in things that kill crops, and the things that kill the things that kill crops) carefully peeling plums before eating them. 'Wouldn't it be easier to just wash those?' I asked. John glanced up. 'If they had live bugs on them, I'd just wash them. Since they don't, I'd rather not risk the stuff they must have on them.'

By the time I moved to Lincolnshire from the West of England, I had been working for Britain's national agricultural advisory organization for three years, during which time I had specialised in cattle, sheep and grassland farming. My job was to provide farmers with advice based on a nationwide network of agricultural research. Initially, the research and advice were all paid for by the government and focused on producing as much food as possible to feed the nation – a legacy of war-time food rationing that had continued until 1954, with meat being the last thing to be de-rationed. But in the 1980s this began to change, as EU food surpluses grew, and environmental management gradually became an integral part of my advisory work with farmers.

Just three years into this well-rehearsed life plan, events suddenly took a strange turn, and I found myself moving to the other side of the country, from the grassy Southwest to the arable East. Colleagues firmly told me there were NO cattle or sheep over there, so I would have to become an agronomist, advising farmers on crop production. At first this seemed to be true. Lincolnshire has some of the best land in Britain – best in terms of arable farming, that is. And, as in most industrialized countries, crop farming had become highly specialized since agrochemicals and sophisticated machinery became available during the mid-twentieth century. Before this, farmers used traditional mixed-ley farming rotations. Arable crops were always followed by grass, grazed by cattle or sheep. The grass and livestock restored the soil after the crops, and they provided a break to reduce the build-up of crop pests and weeds. But decades before I began work, grass and grazing livestock had completely disappeared from crop rotations in eastern England, and most other cropping areas of the world. Every acre that could be ploughed was put down to wheat, oilseeds, potatoes and brassicas year after year. Although livestock remained an essential part of crop rotations on organic farms, these were few and far between; even today, only 1–2% of all arable land is farmed organically in the UK (and most Western countries). The other 98% is farmed with agrochemicals.

However, on the more marginal land to the north and west of the Lincolnshire fens I discovered herds of suckler beef cows – rusty Lincoln Reds, bred by generations of the same family, white-faced Hereford crosses, and increasingly, the newly imported French Limousin and Charolais cattle. There were also a few sheep flocks scattered around, and tens of thousands more sheep arrived from hill farms in Leicestershire and Derbyshire each autumn to clear up thousands of acres of veg waste.

After a year in Lincolnshire, I gladly binned my collection of pesticide catalogues and once again began to enjoy my work with farmers who loved living breathing things, more than steel and chemicals. Standing in a grassy field one day the following spring, surrounded by red cows lazily chewing the cud and tall hedges heavy with white haw blossom, I followed Michael's skyward gaze: the farmer and I both watched spellbound as a hawk hung effortlessly in the sky, and a hare bolted from the tussocky grass. My spray days were over.

One day, I was visiting a farm on the Lincolnshire Wolds. Charles ran a large arable farm, which also had a herd of suckler beef cows on the land where arable crops could not be grown. Straw and other arable by-products were recycled by the cattle, and all their water requirements were met by rainwater collected from the roofs of the farm buildings. Charles was always upbeat and bursting with ideas, but that day he seemed pensive and told me that he and other local farmers had decided to stop using a fungicide called Benlate on their arable crops. A local teacher, Pam Henderson, had begun noticing an unusually high number of children born with eye deformities in rural parts of Lincolnshire, including farmers' children. Then an *Observer* newspaper investigator called Mike Paduano found evidence linking these eye defects to Benlate. The fungicide's manufacturer, DuPont, had hidden the risks for decades. In 1972 they had successfully lobbied against advice from the US Environmental Protection Agency to put a warning label on Benlate. As evidence mounted, Du Pont poured millions of dollars into defending litigation by affected families. Not until 2001 did DuPont finally remove Benlate from the market. Two years later, an American family won a historic case for damages, and that paved the way for action by thirty British families.

When I returned to the West of England after more than a decade in Lincolnshire, my working patch included Dartmoor —nearly 90,000 acres of unenclosed upland moorland. The granite rock, heather moorland and blanket bog had made Dartmoor one of the most well-loved and environmentally important landscapes in Britain, and it had been designated as a national park since 1951. But since then Dartmoor had become one of the most contested of landscapes.

Some environmental activists have described Dartmoor as *sheepwrecked* and some have called for the complete removal of livestock, and the rewilding and reforestation of Dartmoor. These environmental concerns have largely arisen since the 1950s, yet Dartmoor has been managed as common land and grazed with cattle and sheep for 5,000 years. Indeed, livestock grazing helped to create the moorland ecology. So, what changed?

Until the end of the nineteenth century Dartmoor farmers kept the local Red Devon cattle, Dartmoor sheep and Dartmoor ponies. The use of these local breeds (which are less hardy than northern breeds and therefore were not left on the moor over winter) and the traditional *levancy and couchancy* management system (from the French for *rising up and lying down*) ensured that livestock were removed from the moor in winter, and the number of animals grazed there during spring and summer was restricted to the number that *got up and lay down* on each grazier's home farm over winter. Each farm had a limited number of grazing rights linked with it that could not be sold separately. In this way, grazing intensity was kept low and the land was rested appropriately.

From the 1880s onwards some farmers began introducing the hardier Scottish Blackface sheep and Galloway cattle. Change was slow at first, but the Second World War, and impetus for greater food security, led to the 1947 Agriculture Act, which pushed British farmers to increase productivity. More legislation in the 1960s allowed grazing rights to be sold separately from the farms they had been linked with. As a result of these developments, keeping livestock on the moor all year became the norm. And perhaps the most significant change came in the 1970s when Britain joined the European Union (EU) and farmers began to receive headage payments for every cow and sheep they owned. In the next twenty-five years, Dartmoor livestock numbers doubled. As well as this, the traditional practices of cutting bracken and gorse (for thatching and other uses) largely stopped, while the practice of swaling (controlled burning of moorland vegetation to encourage fresh regrowth for grazing) increased. These changes contributed to the spread of inedible bracken, gorse and Molinia grass, increasing the grazing pressure on the remaining grazable areas.

Although Dartmoor (and other British uplands) has undoubtedly been damaged over the last few decades, it is not wrecked. As Adrian Coulson explains in his 2021 study, 'all the necessary building block species are still extant on the moor, albeit at currently reduced populations levels'. Dartmoor can be regenerated, and appropriate livestock grazing has a vital role to play in this, just as it played a vital role in the creation of the moorland ecology for generations before. What matters is how the livestock are managed and the policy making that shapes that management. Grazing with low numbers of animals in the spring and summer helps to prevent deep-rooting woody shrubs from taking over and drying out the peat; so, it supports the growth of the sphagnum moss that builds peat. And livestock grazing provides many other benefits, which we'll discuss further in later chapters, such as preventing wildfires and creating a mosaic of habitat that encourages biodiversity.

Regenerating the ecology of our uplands requires holistic thinking that integrates ecology and farming, not siloed policy making. And success must mean the full involvement of the local people, and the opportunity to earn a decent living too. Hill farm incomes have seen large declines since the 1950s. UK upland Farm Business Incomes averaged £22,800 in 2019/20, despite (or some might argue, because of) four decades in the EU. Hardly rich living for a family business, yet almost all this income was made up by the Basic Payment Scheme (BPS) and environmental schemes. The farming enterprises lost £16,600 per farm. Though beef and lamb are the most nutritionally valuable of all foods, the economics of production are stacked against the farmer. For instance, the 2022 report *Unpicking Food Prices* shows that when a 450g pack of beefburgers is sold for £3.50, the beef farmer gets just 0.1p (0.03%) of the profit, while carrying over 25% of the costs of production. These economics are set to get worse as BPS payments are phased out between 2021 and 2027. The new Environmental Land Management schemes will become almost the only income on these farms.

In these days when most British people live in towns and cities, with countryside access restricted to short trips to a few National Parks, the rewilders may seem like the heroes of this piece, while the livestock graziers may seem the obvious villains and vandals. But the history of British land rights should give us pause. Before the Industrial Revolution, most of us would have lived in rural areas with considerable access to the countryside and its resources. But from the seventeenth century until the First World War, 5,200 parliamentary bills were passed enclosing more than a fifth of English land – 6.8 million acres – removing it from common usage. On Dartmoor, 15,000 acres were removed from common use between 1780 and 1898. The Enclosures were supposed to improve agricultural productivity, but they had the effect of making large numbers of people landless and driving them to the urban mills and mines.

One of the lessons from Dartmoor is that perhaps we should not rush to scapegoat obvious culprits, such as the livestock or the livestock farmers. Many others have played key parts behind the scenes.

Ten years later, I had the opportunity to go and work with pastoral herding communities in Mongolia. On my arrival, I was shocked at the landscape. It resembled the surface of the moon. The exposed soil appeared devoid of plant life. The steppe looked nothing like the pictures I had seen of tumbling grassland and bright wildflowers stretching to the horizon. I couldn't understand what the animals were surviving on. But I was soon to find out. That winter, 8 million Mongolian livestock died, nearly 20% of the national herd. Travelling across the frozen steppe, I saw skinned carcases piled high, the hides being all that the herders had been able to salvage.

This was a dzud; an animal famine. My colleagues blamed it on climate change and overgrazing. Mongolian winter temperatures can drop to -40°C, but this is not a problem for ruminant animals such as cattle. And Mongolian livestock have always lived outside all year round. With thick winter coats and the heat generated by their digestion, ruminants can comfortably survive extreme winters on sparse grazing, provided they begin the winter with plenty of back fat from a summer of good grazing. They naturally draw on this fat through the winter and regain weight during the next summer. One of nature's many cycles. But since 1990 there has been a serious increase in desertification, and 90% of Mongolia's land is now at risk. Why, then, has it become such a problem since 1990? As in the case of Dartmoor, political changes may explain things better than the obvious suspects.

In 1990 communism ended in Mongolia, leading to sweeping socio-political and land-management changes. Much of Mongolia is marginal land covered by steppe grasslands, and about 40% of Mongolia's population still live as pastoral livestock herders. Horses, cattle, sheep, goats, camels and yaks are all reared for meat and milk. The traditional diet was much like the Sapiens diet, with lots of fatty meat, plus dairy products, but few plant foods. However, this has shifted towards the Western diet in recent decades.

Until communism ended, livestock numbers were controlled by the government and grazing management was planned and managed at district level each year. Land and livestock were collectively owned, and herders received a wage from the state. After the end of communism, Western institutions came in and Mongolia was treated to the mantra of *shock therapy* and *rapid transition to a free-market economy.* As my translator, Tamir, told me, 'It was mostly shock and not much therapy.' State-paid salaries ended overnight and there was a free-for-all to grab and breed more livestock. Many herders found themselves with no animals and no livelihoods, leading to unprecedented inequality, poverty and mass migration to urban slums.

In traditional pastoral herding communities, herders keep their animals moving from one area to another, allowing the grassland to rest for long periods before eventually circling back to graze it again. With the sweeping political and economic changes of 1990, Mongolian herders suddenly became livestock owners, who could increase their herd size and move their animals as they wished. But the land remained unfenced and in common ownership. So, herders stopped moving their animals and spread them out to control as much land as possible. Many told me they were afraid to move their animals because other herders would come in and take over the land. This led to continuous grazing, weakening the grassland plants, and encouraging desertification. And it meant that herders could no longer set land aside to make hay for winter feed.

At a crisis meeting in Ulaanbaatar during the 2009 dzud, I was shown a graph of livestock numbers since 1990, which looked like a row of mountain peaks, rising, then collapsing with each dzud. At the time, it seemed obvious to most outsiders that high livestock numbers were the problem. But numbers were not the whole story. I was gradually to understand that the change of grazing pattern and the cessation of herd movement were critical factors. Reflecting on this experience in Mongolia has been another step on my journey towards regenerative livestock farming.

In Mongolia, as elsewhere, political interference and poor policy often lie at the root of mismanagement and apparently natural disasters. The dzuds have continued. Since 2000, they have killed more than 21 million Mongolian livestock.

During the course of this journey, these things have become evident to me:

- The natural world is all about cycles and interconnected living systems (ecosystems); it is a circular economy of sunlight, soil and water, plant life and animals – those that eat plants and those that eat the plant eaters. Nature is not vegan! All these living creatures ultimately return life to the soil. Nature is regenerative.
- Pasture rearing of ruminant livestock can be and generally is the most sustainable form of food production because it works with nature's cycles. Yes, it can be done badly and cause damage – generally when the politicians and consultants get involved. But it can also be done well and produce great environmental benefits as well as great food.
- Intensive livestock farming – housed, grain-fed pigs, poultry, dairy cows and barley-beef – is a completely different matter. It is damaging to the animals and the environment. And one of the main reasons is because it is based on the growing and feeding of arable crops.
- With livestock farming, *it's the how, not the cow*!
- With arable farming, however, *arable farming is the problem*! Crop production is inherently damaging to the environment. It is essentially an extractive process, rather like mining, not a cycle. It depletes the soil of carbon and minerals and erases natural habitat. On its own, arable farming is unsustainable. If we want to continue growing crops, we must stop using agrochemicals and return to mixed-ley farming with grazing livestock in the crop rotations to mitigate the soil damage and replace organic matter

and carbon. Even then arable farming still erases natural habitat, whereas cattle and sheep that are reared regeneratively at pasture can support biodiversity as well as healthy soil.

- The most destructive farming systems are characterized by the absence of livestock. The way we have disconnected arable farming from livestock farming makes crop production completely dependent on agrochemicals. This is unsustainable, and it is responsible for much of our current environmental crisis.

When I began writing about this journey to regenerative livestock farming, I found myself wondering, *is this really where I have arrived – regenerative farming*? Many of the beef and sheep farmers I have worked with over the years have not been regenerative, and I still regard most of them as good farmers who looked after their land and their animals, cared about wildlife and the environment and produced good-quality food. Some have been organic, using no agrochemicals (artificial fertilizer or pesticides). But many, certainly in the early years, were just conventional beef and sheep farmers, and I would argue that even this form of cattle and sheep farming does much less harm and much greater good than arable farming.

These conventional British suckler beef farmers rear their cattle at pasture throughout the summer months. They might use some grains to finish their cattle, but far less than most dairy farmers, or pig and poultry farmers. Some may use a small amount of artificial nitrogen on their grass (many don't), but much less than used by most arable and dairy farmers. It's true that a lot of Britain's lowland beef and sheep pasture today would be described as agriculturally *improved*, the diverse native species having been replaced with a narrow mix of agriculturally productive grasses. But this was mainly done in the mid-twentieth century and most of this land has not been repeatedly ploughed and resown since. So today this beef and sheep pasture is not the kind of sterile monoculture that is seen on many dairy farms, or arable farms. And since it is mostly permanent pasture, the soil and its treasury of carbon are safe from the plough. Even conventional beef and sheep farmers use virtually no pesticides. So, unlike my plant pathologist colleague, I always knew I was safe eating the food those beef farmers produced.

The compromise of conventional (and organic) beef farming that allows some grain feeding means that the farmers can finish their cattle about a year earlier than entirely regenerative herds. In such a low-profit sector, this can make all the difference. But regenerative livestock farming is the ideal. It involves no soil-killing tillage. On the contrary its grazing management nourishes the soil. It uses no ecosystem-destroying agrochemicals; instead its pastures support biodiversity. The question for us consumers, and for our policy makers, is what are we prepared to pay for this?

And now we must talk about BSE (mad cow disease). This national animal disease and public health crisis has forever been linked in the public mind with British beef. BSE was first identified in Britain in the mid-1980s and blighted British farming for the next fifteen years. Countries across the globe banned imports of British beef and beef cattle. Yet, what has never been explained to the public is that BSE was almost entirely a disease of the dairy herd, not the beef herd. Very few suckler beef cattle were affected in Britain or elsewhere. But since around 50% of British beef comes from dairy-bred cattle, all beef was tarred with the same brush. And BSE wasn't just a British issue.

The reason for the lack of BSE in suckler beef herds may be that suckler beef is largely grass-fed, with cows and calves receiving very little supplementary feed. When supplementary feed is used, it is usually home-mixed barley, not the factory-manufactured feed pellets bought in by the majority of dairy farms. It was these manufactured feeds that were implicated as the source of BSE infection (owing to contamination with infected meat and bonemeal). We should also understand that farmers did not knowingly *feed dead animals* to their cows, as some parts of the media liked to make out. Before the BSE crisis, feed manufacturers did not list the ingredients used in their products. All they stated on the bag were the figures for energy, protein, mineral and vitamin contents. Farmers bought the feed in good faith.

Contaminated feed was not the only reason that BSE affected dairy cows rather than suckler beef cows. Genetics and breed susceptibility also played a part. And so did the production stress imposed on the cows by intensive dairy production methods. For instance, a large study of 3,045 French herds found a higher incidence of BSE in the most intensive, high-yielding herds. This study also noted that as in the UK, BSE incidence in France was almost entirely in dairy cows, not beef cattle, and the Holstein breed represented the vast majority of affected cows.

Despite this overwhelming prevalence of BSE in dairy not beef herds, it was the reputation of beef and the livelihoods of beef farmers that paid the heaviest price for the outbreak.

In the next chapter we'll take a closer look at different kinds of farming and the true impacts of our dietary choices on the environment.

CHAPTER 17

NATURE IN CRISIS

Farming and the environment – an introduction

'Wee sleekit, cowrin, tim'rous beastie ...
I'm truly sorry Man's dominion
Has broken Nature's social union,
An' justifies that ill opinion
Which makes thee startle
At me, thy poor, earth-born companion
An' fellow-mortal!'

From 'To a Mouse' by Robert Burns

NATURE IS IN CRISIS. SOILS ARE BEING depleted, wildlife populations have plunged by two thirds in the last fifty years, and a million species of plants and animals are at risk of extinction. We are living through the sixth mass extinction in the planet's history. People blame pollution and climate change among other things, while one of the major causes of this impending cataclysm – arable crop farming – is ignored or glossed over. And to add insult to injury, we have removed the very things that arable farming needs to ameliorate its harms – grassland and grazing livestock.

Crop agrochemicals, ploughing and other forms of tillage, and the continuous farming of arable crops without grass and livestock in the rotations are causing catastrophic wildlife population collapses as well as high greenhouse gas emissions. And the soil itself is dying. Just as Rachel Carson warned in *Silent Spring* six decades ago, crop pesticides trigger a tsunami of destruction, from the tiniest organisms at the bottom of the food chain to the largest birds and mammals at the top, leading to entire ecosystem collapses. Yet sixty years on we have failed to grasp Carson's central message. We have allowed ourselves to be hoodwinked by the agrochemical industry: this new spray will be better, safer and essential to feed the world. We have changed the labels on the poison cans when we should have stopped using the poisons. We are spraying ever-increasing quantities of increasingly toxic poisons over almost all the arable land in the industrialized world, and much of the developing world too.

People who decide to change to a plant-based diet do so with good intentions, believing that it is better for their health, or that it eliminates animal suffering, or reduces environmental harms including climate change. We have already explored the many health benefits of eating meat in Section 2 of this book. In the following chapters we will explore the environmental case for and against livestock and crop farming, beginning with the question of farming and climate change in chapters 18 and 19. And in chapters 20 and 21 we will explore whether plant-based farming really causes fewer animal deaths or less ecological damage than eating meat.

For the authors of this book, the choice to eat meat is one that each of us has made after deep consideration, research, and long experience of animal and crop farming. Central to our decisions are respect for animals and other people; the desire not to cause suffering and harm; the wish to live honestly and in harmony with the natural world, understanding that all food choices have impacts.

This does not mean that we eat any and all meat. It matters how farm animals live and how, when the time comes, they die. What matters is 'the how, not the cow'.

The authors of this book are opposed to the intensive 'factory' farming of animals, as it causes avoidable suffering and environmental harms. For these reasons, we choose to eat meat that is pasture-reared, such as beef and lamb – yes, even though these grazing livestock are ruminants and produce methane. In the following chapters we will explain why this kind of meat is better for the environment than plant foods and hope to convince you likewise. However, we avoid eating meat from intensively reared, grain-fed animals. In industrialized countries this means most pigs and poultry. Common perceptions are that most beef is also intensively reared, but in fact we'll see that it is not.

Many people give up meat and become vegetarian on the basis that it is kinder to animals and the environment. These diets replace meat like beef and lamb with dairy produce and eggs. Yet standard Western systems of dairy farming and egg production have far poorer animal welfare and cause much greater environmental harms than beef and sheep farming.

Dairy farming is not inherently bad for the environment, but food industry economics have driven most dairy producers to become very intensive. Although many British dairy farms still keep cows at pasture in summer, standard management involves high use of grain-based feed, high applications of nitrogen fertilizers, and silage making rather than hay making. Short-term grass leys are generally used, which means ploughing every one to five years, and these leys are generally sown with a narrow range of agricultural grasses. All these aspects of intensive dairy farming reduce biodiversity and increase the risk of soil damage, while slurry from such farms is the most frequent agricultural cause of serious water pollution incidents in the UK. Animal health and welfare issues are also significant concerns in intensive dairy herds, owing to the very high milk yields of modern Holstein dairy cows, and the standard practice of removing calves from the cows a day or so after birth, and rearing these calves on milk replacer and grain-based supplementary feeds. However, there are well-managed, organic dairy farms in the UK as well as a few regenerative dairy producers. And *calf-at-foot* dairies improve animal welfare by leaving calves with their mothers until they reach a natural weaning age. It is up to us whether we are prepared to pay for the benefits these kinds of farms can provide for society.

Eggs and fish are among the most nutritious and healthy foods known to humans and can be a valuable part of the Sapiens diet. However, most egg production in industrialized countries is intensive and grain-fed. Meanwhile, overfishing, and certain fishing and fish farming methods, also cause considerable environmental harms. However, as consumers, we can use and support schemes like the Good Fish Guide, which provides help with sustainable buying choices, while initiatives like Good Catch support sustainable use of fish by restaurants and caterers.

What is it that is so harmful about intensively reared livestock? Animal welfare is one issue. The animals are generally kept indoors in dense groups, which can lead to disease, bullying and other behavioural issues. None of these animal species evolved to eat grains any more than humans did. And just as high-carbohydrate diets cause metabolic diseases in humans, so do the high levels of grains fed to intensive livestock. Water pollution caused by slurry (liquid manure) can be another problem. But a significant reason why we avoid intensively reared meat is the environmental destructiveness of the arable crop farming that it depends on: the grains and soya on which intensive livestock are fed. If one wishes to eat dairy, pork or poultry products, regenerative farms are the ideal source, though not easy to find in the UK currently. Free range, organic dairy, meat and eggs are more available and affordable, and will mitigate many, if not all, of the harms.

At the time of writing, pollution of the River Wye – one of the UK's most beautiful rivers – is attracting a lot of negative media coverage and litigation. The main cause is manure leaching from intensive poultry units, as well as human sewage. The criticism of these intensive farms is justified, and we join with it. This should not be happening, especially as pollution from livestock waste is entirely preventable, with the right farming systems and management.

However, the outspoken critics of these polluting livestock units rarely mention the major environmental harms that are also caused by crop farming – harm that is caused by the majority of crop-farming activities, not just a part of the sector, as is the case for livestock. Despite the spotlight on poultry pollution of the River Wye, when we consider UK farming as a whole, arable farming is the biggest polluter after intensive dairying, causing more water pollution than poultry and pig farming. Meanwhile, beef and sheep are the least polluting of all UK farming systems, but when did you ever read that in the *Guardian*? And when the media runs a story on pesticides or the catastrophic collapses of insect and bird populations that we are witnessing worldwide, they conveniently forget to mention that virtually all these pesticides come from arable crop farming. Almost none are used in beef and sheep farming. Nearly all cropland in the industrialized world is treated with these ecosystem killers, while hardly any grassland is. We'll look at this more closely in chapter 20.

As with medicine and most other sectors, farming has its own jargon and assumed knowledge. So, the Farming 101 section below describes some farming basics and definitions to help us understand the following chapters without disrupting their flow.

FARMING 101

Firstly, it's worth knowing that farming is a very recent activity. People began doing it only 10,000 years ago. This is a blink of the eye compared with the previous 300,000 years that we *Homo sapiens* lived by hunting and eating meat.

Over the course of these last 10,000 years, humans have domesticated plants for crop farming and animals for livestock farming. And we have changed both very significantly compared with their wild ancestors, by selectively breeding for the characteristics we want, such as grain size and yield, or the amount of muscle or milk yield. But there is a critical difference here: humans always ate meat, and livestock domestication has enabled us to continue doing so without negative impacts on our health. By contrast, humans ate very few plant foods during our evolution. The domestication of plants radically changed them, making them much more palatable than their wild ancestors. As a result, most of us are now eating large amounts and types of plant foods that humans did not evolve to eat. As discussed in the early chapters of this book, this is at the root of many of our health problems.

Around 1.6 billion hectares (12%) of the world's land area is used to grow arable crops. These crops are mostly grown as monocultures (just one type of plant covering a large area), something that is not seen anywhere in nature.

Most crops are annual plants that have to be planted each year (the exceptions to this are perennial fruit and nut trees). This generally involves the complete removal and replacement of natural vegetation, doing untold damage to natural ecosystems in the process.

The plough was invented only 4,000 years ago. It revolutionized crop farming and is often celebrated as one of humankind's greatest inventions. But now we are beginning to recognize its significant environmental downsides. Ploughing and other forms of tillage expose the soil's organic matter to the air, so that its carbon content is converted to carbon dioxide (CO_2) and released into the atmosphere. Until people began tilling the ground to sow crops, the Earth's soil was almost always covered with diverse wild plant life, including large areas of natural grasslands; the soil was undisturbed and rarely exposed to the air. But today we till around 1.4 billion hectares of land worldwide every year, constantly releasing carbon from the soil to the atmosphere. Around a third of the CO_2 that human activity has added to the atmosphere has come from soil damaged in this way. Yet the predominant media narrative is that it is methane from cow burps (yes, its burps not farts!) that causes climate change.

Livestock farming means rearing domesticated animals, such as cows, sheep, pigs and chickens, for food, fibre (such as wool) and hides, or to pull ploughs and carts. Grazing livestock such as cows and sheep can be reared entirely at pasture, whereas non-grazing livestock like pigs and poultry are generally reared on grains and soya (though these are not their natural diet).

Unlike arable crops, grasses and other plants found in grasslands are perennial plants that grow continuously and are adapted to being grazed by herbivores. So, pasture on which livestock are grazed never needs to be ploughed. In many parts of the world pastoralists and farmers graze their livestock on natural or semi-natural grasslands rather than replacing the natural habitat as crop farming does. However, it must be acknowledged that during the mid-twentieth century many areas of natural or semi-natural grasslands were significantly damaged by agricultural *improvement* to increase livestock productivity (reseeding with agricultural grasses, drainage, fertilizer and herbicide application), as well as for conversion to arable farming. In Britain after the

Second World War and the 1947 Agriculture Act, thousands of acres of lowland species-rich meadows, as well as some upland moorlands, fell victim to these *improvements*. Indeed, my first job after university was as a researcher on a government-run research farm situated high on Exmoor that before my time had pioneered some of these *improvements*. Approaching the 450-acre farm across the moor, it resembled an emerald of ryegrass, a bright green blob in the middle of the softer, mottled tones of the surrounding moorland.

Though most of the *improved* land in the lowlands remains relatively species-poor today, beef and sheep grazing is the least intensive form of land use, rarely ploughed and with little agrochemical use; not perfect, but far less damaging than the arable farming that will have to replace it if we are all to become more plant-based. Meanwhile, moorland agricultural *improvements* (have we made it clear what we think of these improvements?) stopped in the 1980s, and some areas have been restored. But of course we can and should do more, in the lowlands and the uplands. That's where regenerative livestock farming comes in.

Badly managed livestock can contribute to soil damage and carbon loss, but with good grazing management they are beneficial for the soil and carbon capture in the soil, as we'll see when we look at the biogenic carbon cycle in chapter 18.

The livestock we are most concerned with in this book are ruminants, such as cows and sheep, because the misconceptions surrounding ruminants, climate change and their other environmental impacts need putting straight. Ruminants are herbivorous mammals that evolved 50 million years ago, and there are about 200 species today. About ten of these species have been domesticated for livestock farming, including cattle, sheep and goats. The rest are wild ruminants, such as giraffe, antelope and wildebeest.

What makes ruminants so important is their specialized digestive system. Their large stomach is divided into four chambers, the first being the rumen (from which they get their name). The rumen, and the billions of microbes it holds, act like a fermentation vat and enable the ruminant to digest plant fibre more effectively than can any other kinds of animal, to extract maximum nutrition, even from poor-quality vegetation. The rumen microbes provide all the energy and protein the ruminant needs. This has allowed ruminants to thrive throughout most of the world, from the tropics to the Arctic Circle. The issue here is that the rumen microbes produce methane, which the cow or sheep burps out into the atmosphere.

We humans have one simple stomach and cannot digest plant fibre at all. For at least the last 300,000 years of human evolution our survival has depended on large herbivores, especially ruminants, to turn low-quality vegetation into nutrient-dense human food – meat. This enabled us to disperse around the world too. Our hominin ancestors came out of the forests millions of years ago to evolve in Africa's grasslands, where meat but very little fruit or other edible plant food was available. Remember that the wild ancestors of the plants we eat today were largely inedible. Later, humans followed the herds out of Africa into Eurasia and the cold north. Even today, two thirds of the world's land that provides food for humans does so only through grazing animals, because this land is suitable only for grassland, not arable crops.

Pigs and poultry are not ruminants. Like us, they have one simple stomach and cannot digest fibre effectively. So, whereas cows and sheep can obtain all the nutrition they need from fibrous grassland plants, with no need for starchy grains or extra protein, pigs and chickens cannot. In industrialized countries virtually all poultry and pigs are reared intensively on grains, from arable land. Some would argue this land would be better used to grow crops for direct human consumption.

Pigs and poultry also require extra protein, often supplied as soya bean meal. Soya is widely grown in the USA, so US livestock farmers feed US soya. But soya is not grown in the UK and almost all UK imports come from the Amazon where they contribute to deforestation. Of these UK soya imports, 35% go into pig and poultry feed. But pigs and poultry are not the worst culprit in this regard. The largest proportion of British soya imports (45%) is used to manufacture alternatives to dairy and meat, for plant-based consumers. Despite common misconceptions, less than 4% of British soya imports are used in ruminant feeds (most of that is fed to intensive dairy cows, not beef cattle or sheep), because the ruminant can obtain all the protein it needs from pasture, thanks to its unique digestive system.

Another reason why we do not advocate intensively farmed livestock such as pigs, poultry, barley-beef and intensive dairying is their poor reputation for causing water and air pollution. This is mainly because these intensive systems tend to produce large volumes of slurry (liquid manure), rather than the less harmful solid manure that extensive beef and sheep farms generally produce. However, even pollution from intensive livestock units is entirely preventable with good design, maintenance and management.

Most of the world's grasslands, where beef cattle, sheep and other grazing livestock are reared, is marginal land, meaning that it is unsuitable for growing arable crops. The soil is too poor, the growing season too short, the climate too cold, too wet or too dry, the slope too steep, and so on. Two thirds of the land used for food production globally is marginal land, including more than half of British farmland. If grazing livestock such as cattle and sheep were not reared here, the land would not produce any food. Indeed most ruminant livestock are

Sheep grazing in the Shropshire Hills, UK. Grazing livestock produce food from marginal land that is unsuitable for arable crops.

reared on marginal land, not arable land. These pasture-reared ruminants provide extra food for the global table, and do not steal land that could be used to grow crops for human consumption.

However, grass and grazing livestock have traditionally played an essential role on arable land too. In traditional mixed-ley farming, arable crops were followed with grass, grazed by livestock, to restore soil organic matter and soil nutrients depleted by the arable crop. This also provided a break to reduce crop diseases and weeds. This is still the case in organic and regenerative crop production today, and in many parts of the developing world. But over the last century, conventional crop farming in industrialized countries has become increasingly disconnected from livestock. Instead, arable crops are grown continuously. This means heavy reliance on artificial fertilizers and pesticides. If we wish to continue growing arable crops, and stop the devastation they are causing to our soils and ecosystems, we must return to regenerative mixed-ley farming with livestock.

Beef production is often criticized for high consumption of grains, and hence its use of food or land that could provide plant foods for humans. But 90% of the world's beef, including most British beef, is produced in low-intensity pasture-based systems. Figures commonly cited for the amount of grain fed to produce 1kg of beef (typically 6–20kg of grains) are highly skewed, as they are often based on feedlot (barley-beef) production which accounts for only 10% of global beef production. Even in the USA where feedlot systems are most common, beef production is predominantly pasture-based, with cattle moving to a feedlot only for the last weeks of fattening. And these grain-feeding figures also obscure the fact that 40–70% of the feed used is actually not edible for humans, but inedible food industry by-products such as brewers' grains, maize gluten, and oilseed meals.

Sheep farming in Britain and globally tends to be extensive and pasture-reared with very little grain feeding or agrochemical use.

A FEW OTHER FARMING TERMS

Arable land: Land that can be cultivated and used to grow annual arable crops, such as grains and vegetables.

Marginal land: Land that is unsuitable for arable crops and can generally only be used for grazing or trees.

Semi-natural grassland: Areas that exist as grasslands (with few trees or woody shrubs), as a result of human activity such as grazing or mowing. However, most of the plant species present are native grasses and broadleaved species not agricultural species. These grasslands are species-rich and ecologically valuable. And it's worth remembering that animals also change and adapt their environment for their own benefit – beavers, for example.

Pastoralism: Pastoralists live by herding grazing livestock across range (unenclosed) grasslands, often moving nomadically across large distances in marginal grazing lands such as deserts, the central Asian steppes and the African savannah.

Organic livestock: Organic farming allows the feeding of grains to livestock, but the grain are grown without agrochemicals (pesticides or artificial fertilizers). And no artificial fertilizers or pesticides are applied to pasture.

Regenerative livestock farming: This involves no grain feeding or agrochemicals.

Holistic planned grazing: Throughout my working life farmers have argued the pros and cons of systematically moving the herd around a series of paddocks (rotational grazing), or spreading the herd out over a wide area without moving it (set-stocking). In recent years there has been great interest (especially among regenerative farmers) in a type of rotational grazing known as 'mob grazing' or holistic planned grazing, as developed by Allan Savory. We will look at this further in chapter 19.

Beef farming: This means the rearing of cattle for meat as opposed to milk (dairy farming). The cattle are usually specialist beef breeds rather than dairy breeds. Selective breeding for milk yield opposes selection for muscle development and vice versa. This is why dairy cow breeds like the Holstein look much skinnier than beef breeds like the Aberdeen Angus; and why beef breeds are not generally used for milking, although dual-purpose breeds are common in the developing world just as they once were in Britain.

Suckler cow/suckler beef cow: In Britain about half of our beef comes from suckler beef herds. They are called sucklers because the calves remain with their mothers, suckling the cows for six months or more (at pasture), rather than being separated from the cow within days as most dairy calves are. The bulls come from beef breeds. Typically on hill farms, the cows will be pure native beef breeds such as the Welsh Black or Scottish Galloway. Whereas on lowland suckler farms, the cows are often a beef breed crossed with a dairy breed. Until the 1980s the most common lowland suckler cow was the Hereford x Friesian, produced by dairy farmers using a Hereford bull on their Friesian dairy cows. Suckler beef is the best-quality beef owing to the beefy muscling of beef breeds, and the way the calves are reared with their mothers at grass. The meat is often sold through traditional independent butchers' shops.

Dairy-bred beef: Dairy cows must have a calf before they can produce milk. A dairy farmer would typically mate around half the cows to a dairy bull, and half to a beef bull breed. Hence around a quarter of the calves would be pure dairy heifers (females) to provide replacements for older cows leaving the herd; a quarter would be pure dairy male calves that would be reared for beef; a quarter would be beef x dairy male calves that would also be reared for beef; and a quarter would be beef x dairy female calves that could become suckler cows. Though the meat from dairy breeds is just as nutritious and healthy as meat from beef breeds, dairy breeds are much less well muscled, so they produce less meat in the most valuable parts of the carcase such as the sirloin and rump. So, dairy-bred beef is often sold through supermarkets, as mince or in processed products such as burgers and sausages.

So, now that we have got our welly boots firmly on, we're ready to jump straight into the Bullsh*t and look at that central pillar of the case against meat, the question of cows and climate change.

CHAPTER 18

BULLSH*T

Cows, crops, the climate and creative accounting

'The most difficult subjects can be explained to the most slow-witted man if he has not formed any idea of them already; but the simplest thing cannot be made clear to the most intelligent man if he is firmly persuaded that he knows already, without a shadow of doubt, what is laid before him.'

Leo Tolstoy

HARDLY A DAY GOES BY WITHOUT MEDIA headlines on climate change, together with calls for action. The solutions seem clear so we should just get on with it. And when it comes to the role of diet in climate change, common perceptions are that meat is bad, whereas plant foods are good. So we should stop beefing about it and eat more beans – thrrrrrppp! Pardon me!

Furthermore, the messaging and perceptions are that beef is the worst of all meats because cows produce methane, and that methane is said to cause climate change. But is it really so simple? Meat and livestock farming play critical roles in human nutrition, global food and fibre supply, and human culture. And when it comes to the climate impacts of farming and food production (crops and livestock), there is less agreement than you might think.

In this chapter we will look at the climate case against livestock farming. Are all forms of livestock farming harmful, and are cows really the worst? What about crop farming? What does the United Nations (UN) say? What do others say? And how reliable is the information on which many of us base our views? What unforeseen consequences of such farming changes might we be overlooking?

A recent conversation illustrates the distortions and misunderstandings we commonly hear about cows and climate change. I was visiting some friends, whom I knew to be concerned about the environment. I invited them to an open day at a local regenerative farm. Gerry passed me a cup of coffee as we talked, but I waved away the soya milk he offered, surprised that he was not concerned about the environmental impacts of soya growing. I continued explaining how the farm's herd of beef cows played a central part in its environmental management and climate change mitigation. Nevertheless, Gerry refused my invitation, stating that he was vegetarian and 'everyone knows that beef cattle are the biggest cause of climate change!' This extraordinary claim couldn't be further from the truth, but Gerry's distorted belief about cows is common.

In Britain and most Western nations, only around 1% of us work in farming these days, down from 23% in the mid-nineteenth century. We have lost connection with and understanding of how our food is produced. Meanwhile few of us have the time or expertise to grapple with UN climate change reports. So, most people accept what they hear in the media, such as in the movie *Cowspiracy*, or the *World Watch* magazine article 'Livestock and Climate Change' by Robert Goodland and Jeff Anhang, on which the movie was based. The article was not peer-reviewed and it was based on flawed and widely criticized calculations. The movie and the article claimed that livestock farming was responsible for 51% of man-made greenhouse gas emissions. As we'll see in the section below, this figure is more than four times the latest estimates of livestock emissions published by the UN's Food and Agriculture Organization (FAO) or by the Intergovernmental Panel on Climate Change (IPCC). And for reasons we'll go on to discuss, even these much lower UN figures overstate livestock emissions.

EMISSIONS CRITICAL

Climate change is a big subject. It is not within the scope of this book to discuss all its aspects, except to mention that many natural phenomena such as solar cycles, plate tectonics, wind and ocean circulations, and clouds play their parts in the Earth's climate. Here we must focus on those aspects of climate that pertain to farming and food systems, particularly greenhouse gas emissions.

Greenhouse gases currently make up about 0.042% of the atmosphere. The remainder being nitrogen (78.08%), oxygen (20.95%) and argon (0.93%), which do not contribute to global warming. The atmosphere also contains up to 4% water vapour (40,000 parts per million). The greenhouse gases that the IPCC focuses on are: carbon dioxide (CO_2) – 0.0425% of the atmosphere or 425ppm; methane (CH4) – 0.000193% or 1.93ppm; and nitrous oxide (N2O) – 0.00003% or 0.3ppm.

Despite being a small proportion of the atmosphere, greenhouse gases cause a significant warming effect by trapping some of the sun's radiation. This has benefits, as without any of these gases the Earth would be much colder. However, since the mid-eighteenth century atmospheric CO_2 levels have risen by 50%, CH4 by 150% and N2O by 20%. The IPCC regards these increases as the cause of global temperature rises and climate change. Although there are many natural biological and geological processes that produce greenhouse gases, these increases coincide with the industrial era, and the IPCC considers them to be the result of human activities, particularly the burning of fossil fuels.

While most of this chapter is about methane, because of its specific relevance to ruminant livestock, we also need to mention water vapour (H2O) here. If you look again at the atmospheric gas levels listed above, you'll notice that they add up to 104% when water vapour is included. That is because climate bodies normally exclude water vapour and cite the other gas levels as percentages of the dry atmosphere, although in reality humidity never drops to zero, even in the hottest deserts.

Water vapour normally comprises up to 4% of the atmosphere. That is a hundred times higher than the level of CO_2 and 22,000 times higher than the level of methane; and the IPCC recognizes that water vapour is a much more potent greenhouse gas than carbon dioxide, methane or nitrous oxide. The IPCC does not regard water vapour as a cause of climate change, however, because it mostly comes from natural sources; and although water vapour has a warming effect, the IPCC view is that its contribution to climate change is a secondary effect of temperature changes caused by man-made emissions of the other greenhouse gases. This aside, water vapour is very important to our discussion of farming's impacts on climate and ecosystems, because different land uses, and changes in land use – such as deforestation and the conversion of grassland to arable crops – change the amount of water vapour and cloud cover in a locality. This affects the local climate, temperature, habitat and ecosystem. We will come back to this in chapter 19.

So far we have only talked about emissions, because these are what the IPCC and FAO focus on. However, net atmospheric levels of greenhouse gases are also affected by processes that remove them from the atmosphere. The Earth's natural climate regulation involves many feedback systems and cycles, such as the water cycle and the carbon cycles (fast and slow), though the IPCC does not usually take much account of these. Both CO_2 and CH4 contain carbon, and, as we shall see later in this chapter, grasslands and grazing animals like cows, play particularly important roles in the biogenic (fast) carbon cycle, drawing down carbon from the atmosphere and storing it in the soil (carbon sequestration).

According to the IPCC, fossil fuel burning (for energy, transport and industry) is by far the biggest source of man-made greenhouse gas emissions (76%). Meanwhile, the latest in a series of global livestock emissions assessments published by the FAO puts emissions from global livestock farming at under 12% of all man-made emissions. Although this figure is only a fraction of the wildly inflated 51% *Cowspiracy* claim, these FAO assessments still overstate livestock emissions, especially those from grazing ruminants. Before looking at these assessments more closely, perhaps we should take a brief look at crop-farming emissions.

You'd be forgiven for wondering, *Do crops produce greenhouse gases? Well, I never!* We rarely hear anything about crop-farming emissions in the media. Plants are green, and green is good, isn't it? But we have to eat something. And calls to reduce or end livestock farming assume that animal foods will be replaced by growing more crops. But what if crop farming produces as many or more emissions, and causes other harms? This failure to properly consider the other side of the coin is a major problem. The 2018 meta-analysis 'Reducing food's environmental impacts through producers and consumers' by Poore and Nemecek is a typical example of this failure, yet it is frequently cited in the media as justification for cutting animal foods from our diet. As a meta-analysis, the study reviews other studies, like the FAO livestock assessments, so it simply echoes and ingrains the same faults, and has also been widely criticized.

The FAO global livestock emissions assessments include all emissions from farm to fork. Yet it is hard to find comparable assessments of global crop farming. The figures that are available often exclude significant emissions from the crop farming balance sheet.

To begin with, crop farming involves a lot of mechanical operations like tilling, spraying and harvesting. These burn a lot of diesel, producing lots of emissions, but UN analyses generally allocate these emissions against the energy sector, not the crop-farming sector. Transporting crops to market also produces emissions, but UN analyses allocate these against the transport sector.

Most crop growing involves tilling the soil every year, which, as explained in the previous chapter, releases large amounts of CO_2 to the atmosphere, but this is not accounted for in the UN analyses or Poore and Nemecek's study. In fact, around a third of the increase in atmospheric CO_2 originates from such soil damage, but this is not widely reported.

Most fertilizers and virtually all pesticides are applied to crops, not grassland (grazing livestock), but the very significant emissions from their manufacture and application are rarely counted against crop farming. We will take a closer look at crop farming's wider environmental impacts in chapters 20 and 21.

To be fair, intensive livestock farming of pigs, poultry, dairy cattle and barley-beef also produce a lot of fossil fuel emissions. Most intensive livestock are reared in housed systems that require mechanized feeding, watering and waste management. Pig and poultry housing may require heating in winter and mechanical ventilation in summer. And growing grains and soya to feed these livestock is responsible for a lot of emissions. But FAO global assessments treat these emissions in the opposite way to those from crops, allocating them to the livestock sector, not the energy sector. These emissions are one of the reasons we are not in favour of intensive livestock farming, any more than intensive crop farming. However, the difference is that almost all crop farming in industrialized countries is intensive with high inputs and emissions, whereas most of the world's beef and lamb is reared in pasture-based systems with the lowest inputs of fertilizers, pesticides and fossil fuels of any food production system. Ruminants do not require heating or mechanical ventilation, even when housed during winter. And since grassland for beef cattle and sheep is rarely ploughed, there are few if any emissions from its soil.

THE CLIMATE CASE AGAINST LIVESTOCK

The FAO has produced estimates of global livestock emissions beginning at 18% in 2007, but revised down to 12% in its most recent review. Despite downwards revisions and continued scientific criticism of how these assessments exaggerate livestock emissions, dramatic media headlines such as 'cattle produce more greenhouse gases than cars' have stuck in the public mind.

One way that the FAO reports manage to make the livestock sector look bad is by comparing it with the transport sector, and cherry-picking what data they include. Many transport sector emissions are left out, such as those from vehicle construction and the mining of metals; only direct vehicle exhaust emissions are compared, while all livestock farming emissions are counted. In contrast to the FAO assessments, IPCC comparisons (which include all relevant emissions) have found that the transport sector produces two to three times more emissions than livestock farming.

In the USA, where meat consumption per capita is the highest in the world, agriculture is reported to be responsible for 9% of total US emissions, approximately equal amounts from crops and livestock (though net livestock emissions are lower after accounting for carbon sequestration by pasture-reared animals, which we'll discuss below). Meanwhile, transportation is responsible for 29% of US emissions and industry for 22%. Similarly, in the UK agricultural emissions are reported to be 10% of total UK emissions, again with approximately equal contributions from crops and livestock, while transport produces 27%, energy supply 24%, business 17% and residential 15%. So, should cutting meat consumption be the priority for US and UK consumers?

We have to eat something. Yet neither the FAO nor the IPCC provide comparisons of crop sector emissions against livestock farming (nor crop sector emissions against the transport sector), perhaps because crop sector emissions are worse than pasture-reared livestock, as we'll show in the following chapters.

Another problem with the FAO assessments, and Poore and Nemecek's study, is the way that they lump all the varied kinds of livestock-rearing systems together as one global *livestock sector*. Professor Ian Scoones of the University of Sussex, a specialist in pastoralism, who leads the international PASTRES programme, has summed up the *livestock sector* problem like this: 'The notion of a "livestock sector" presented in many global assessment reports is largely meaningless. There are hugely different livestock production systems in different parts of the world: from contained, industrial factory farming to extensive grazing on open rangelands. These different systems have very different emission dynamics.'

UN climate bodies and media ignore this diversity in livestock farming. And the environmental benefits that livestock farming can and does bring are also ignored. In the rush to vilify meat and promote plant-based or laboratory-grown foods, media and lobby groups constantly present an image of sick, factory-farmed animals, crowded in sheds, pumped full of grains and antibiotics, dirty, smelly and polluting. We agree that these intensive livestock farming systems harm the environment and cause suffering to the animals. But when it comes to beef cattle and sheep, this factory-farming image is far from representative.

Less than 10% of the world's beef cattle are intensively reared on grains in feedlots ('barley-beef'). This kind of rearing system is mostly seen in the USA and Latin America. The vast majority of the world's beef cattle and sheep are reared at pasture, including most British beef and sheep. These pasture-reared livestock farms have the lowest use of artificial fertilizers of any food production system, nearly zero pesticide use and the smallest water footprints of any farming system (see chapter 21); and the lowest use of grain feeding, antibiotics and other veterinary medicines of any kind of livestock farming. UK and EU livestock are completely free of growth-promoting hormones (those were banned in the 1980s). And the bonus is that all pasture-reared livestock add organic matter and carbon to the soil. This not only improves soil fertility and avoids or reduces the use of artificial fertilizers, but it also mitigates climate change.

The FAO assessments give the misleading impression that pastoral herders, such as the Maasai rearing cattle on African grasslands, and British farmers rearing cattle on grassy hillsides, feed their animals just as much grain and use just as much fertilizer as those rearing barley-beef in feedlots, and that they are just as responsible for deforestation in the Amazon as the Amazonian beef ranchers. This is like saying that the population of the UK eats a 1,000 tonnes of beef each year and vegetarians make up about 10% of the population, so British vegetarians eat 100 tonnes of beef each year.

THE CLIMATE CASE AGAINST COWS

Let's now take a closer look at ruminant livestock specifically, such as cows, sheep and goats. The FAO assessments especially highlight emissions from beef cattle. UN emissions estimates for the global dairy sector are similar to those of the beef sector, but in fact the way most dairy cows are farmed in industrialized countries is far more intensive and environmentally damaging than beef farming. A further point to bear in mind is that in many parts of the developing world, cows are multi-purpose animals, providing milk and meat, and pulling ploughs or carts. In Britain, there is also a large degree of integration between the dairy and beef herds, with about three-quarters of the calves from the dairy herd transferring to beef herds to grow on, or to become breeding beef cows. If we all agreed with vegetarians that it is fine to eat dairy produce but not beef (and it is certainly necessary to eat animal foods of some kind), all those dairy calves would have to be slaughtered at birth.

The FAO assessments indicate that beef and dairy farming each produce 3–4% of global greenhouse gas emissions, mainly from methane and feed production emissions. But as we will now see, these emissions are exaggerated. And the assessments fail to account for the significant ways in which pasture-reared livestock reduce emissions by recycling crop wastes and by drawing carbon from the atmosphere into the soil (carbon sequestration). And even if these emission levels were correct, what are the trade-offs in terms of the major contribution that pasture-reared livestock make to human nutrition and the global food supply? Would eliminating ruminant farming really reduce overall food emissions or would they be replaced with more emissions from more crop farming?

METHANE – THE MOST MISUNDERSTOOD EMISSION

The methane that ruminants produce during digestion, and belch out, accounts for 43% of beef and dairy farming emissions, according to the FAO assessments. But is ruminant methane really the problem it's made out to be? Does it increase the net amount of methane or carbon in the atmosphere, increasing global warming, or are there other factors at play?

Professor Myles Allen, who leads Oxford's Climate Dynamics Group, and has authored several IPCC reports, says, 'even if we were to eliminate methane emissions from the entire ruminant herd worldwide, we would [only] shave a few hundredths of a degree off global temperatures ... that is [the equivalent of] a few years of fossil-fuelled warming'.

The IPCC says that atmospheric levels of methane have increased by about 150% during the last 200 years. But, while ruminants certainly emit methane, so do many other human activities, including the burning of fossil fuels, poor management of landfill sites and waste, and rice growing. Interestingly, we rarely hear calls to end rice growing or rice consumption.

So far, the scientific community has not been able to identify which of these activities are causing the increase in atmospheric methane levels. But remember that ruminant methane is a flux gas. It is part of the biogenic (fast) carbon cycle, which takes only seven to twelve years to complete (see, The Biogenic Carbon Cycle, below). The methane that ruminants belch out originates from atmospheric CO_2 via the plants that the ruminants eat. It breaks down to CO_2 again in the atmosphere, replacing the CO_2 the plants removed, though not all of it. Ruminant livestock, like wild ruminants, are just part of this rapid cycling of methane and CO_2. No new methane, CO_2 or carbon is created in the biogenic carbon cycle, so methane from ruminants does not accumulate.

However, the carbon emitted as methane or CO_2 when we burn fossil fuels comes from ancient coal, oil and gas stores underground, not from the atmosphere. It took billions of years for the slow carbon cycle to deposit that carbon as fossil fuels in rocks. When we burn fossil fuels, huge quantities of this carbon are rapidly released into the atmosphere as CO_2 and methane. The slow carbon cycle is too slow to remove these quantities of fossil fuel methane and CO_2 from the atmosphere back underground, so they are effectively new methane and CO_2. Although methane breaks down rapidly in the atmosphere, CO_2 does not break down easily. Unless it is removed by plants and algae for photosynthesis CO_2 can linger in the atmosphere for hundreds or thousands of years. So, the rapid production of extra methane and CO_2 caused by burning fossil fuels causes these gases to build up in the atmosphere.

Ruminants have been on Earth, emitting methane, for 50 million years, and not in small numbers. In prehistoric times the total number of wild ruminants was as high, and sometimes higher, than the total number of wild and farmed ruminants we have today. And 200 years ago, the total number of ruminants was similar to today. When the first European settlers arrived in North America there were 60 million buffalo roaming the plains, as well as huge herds of elk and other wild ruminants. So how can ruminant methane emissions explain increases that have occurred only in the last 200 years? And if ruminants were really responsible for the recent increases, one would expect atmospheric methane levels to be highest in those parts of the world where ruminant numbers are highest, and vice versa. But this is not the case.

THE BIOGENIC CARBON CYCLE

Ruminants emit methane but they do not add to the overall level of methane or carbon in the atmosphere. This is because they are part of the biogenic (fast) carbon cycle (below). They contribute to carbon sequestration processes that remove carbon from the atmosphere and transfer some of it to the soil. So grasslands and grazing animals produce a net transfer of carbon from the atmosphere to the soil. This mitigates atmospheric warming, rather than contributing to it.

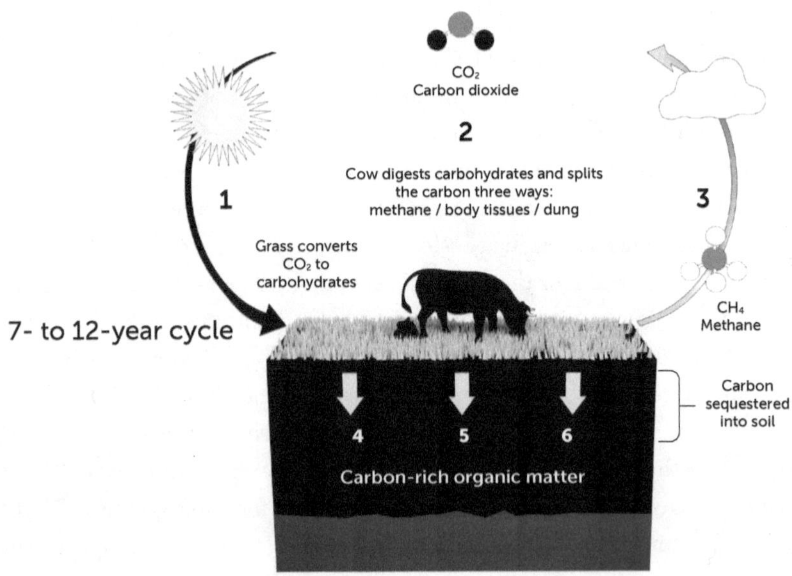

The biogenic cycle of carbon in the atmosphere and soil.

1. Grassland plants extract carbon dioxide (CO_2) from the atmosphere during photosynthesis and convert it to various kinds of carbohydrate ($C_6H_{12}O_6$), including sugars, starch and cellulose.

2. The cow eats some of the plant carbohydrate, containing some (but not all) of the carbon that came from the atmospheric CO_2. The cow splits this carbon in three directions:
   - Some carbon (but not all) is converted to methane (CH4) and the cow belches it out into the atmosphere,
   - Some carbon is incorporated in the cow's body tissues,
   - and some is excreted in the dung.

3. In the atmosphere, the methane is broken down into CO_2 again. This replaces some (but not all) of the CO_2 that the grass removed from the atmosphere.

   There is no increase in atmospheric methane or CO_2 because this is a continuous flux of carbon in and out of the atmosphere through the plants and animals. No new methane or CO_2 is created. The biogenic carbon cycle is known as the fast cycle because it takes only seven to twelve years for the methane to break down to CO_2. As described earlier when we discussed the slow carbon cycle and fossil fuels, CO_2 does not break down easily in the atmosphere. But this biogenic CO_2 is absorbed by plants again and removed within the same seven- to twelve-year cycle.

   In fact, the grassland and grazing animals cause a net reduction in atmospheric carbon (methane and CO_2) because both the grassland plants and the grazing animals divert some of the carbon elsewhere. Some is incorporated into their body tissues. And some is diverted into the soil by various carbon sequestration processes which increase the amount of carbon stored in the soil as organic matter.

Carbon sequestration processes:

4. The roots of grassland plants exude carbon-rich sugar ($C_6H_{12}O_6$) into the soil. Fungal mycelia surrounding the roots provide an exchange medium, allowing the plant roots to trade sugar across to soil organisms, in exchange for micronutrients that the plant needs. Grazing animals have been shown to stimulate grass roots to secrete more sugar into the soil, and to stimulate soil microorganisms that increase sequestration through other processes.

5. Carbon is also sequestered into soil via the cow dung falling on pasture (or spread as manure). This carbon-rich dung is rapidly incorporated into the soil by dung beetles, worms and other soil organisms (the livestock also support biodiversity in this way).

6. Dead plant and animal tissues falling on the soil surface provide another carbon sequestration pathway, as the organic matter is drawn down and incorporated into the soil by soil organisms.

Beef cattle and sheep are generally reared on unploughed (permanent) grassland, using few if any agrochemicals. The grass provides almost complete soil coverage and fills the soil with its roots, while the dense fungal network surrounding the roots remains undisturbed. So, carbon sequestration is optimized in such grassland soils.

In these ways, carbon from atmospheric CO_2 is drawn down via plants and animals and stored in soils, especially grassland soils. Grasslands soils hold and sequester more carbon than temperate forest soils, and grasslands store carbon safely underground, whereas forests store a lot of their carbon in the tree biomass above ground, where it is vulnerable to fire. But carbon sequestration and its importance in mitigating emissions are generally ignored in UN emissions assessments.

By contrast, arable crops are very poor at sequestering carbon into soil; in fact, they do the opposite. For a start, ploughing and other forms of tillage, and the agrochemicals routinely used in crop farming, destroy the soil fungal network that helps carbon sequestration. Secondly, arable crops provide much poorer coverage of the soil surface than grassland. They cover the land for only part of the year, and they are thinly spaced compared with grassland plants. This, along with tilling, means that a lot of the soil's organic matter is exposed to the air a lot of the time. The oxygen in the air reacts with the carbon in that organic matter, releasing CO_2 into the atmosphere. This not only creates CO_2 emissions, it also depletes the soil of the organic matter and carbon that make it fertile and which hold water. This in turn makes the soil more vulnerable to erosion and desertification.

GUESS HOW MUCH METHANE WARMS?

Just as scientists don't fully understand where the additional atmospheric methane has come from, nor do they understand how much warming methane actually causes. The IPCC uses a methodology called global warming potential

(GWP) to calculate the warming impacts of greenhouse gases. But even IPCC scientists say that GWP overestimates the warming impact of methane. Theoretically, a tonne of methane has a larger warming impact than a tonne of CO_2 at the instant each is emitted to the atmosphere. But the methane decomposes within about ten years, rather than hanging around in the atmosphere for a thousand years like CO_2. So, methane emissions don't accumulate in the atmosphere as CO_2 emissions do.

Since the climate change case against ruminants is largely based on their methane emissions, the inaccuracy of the GWP methodology is one of the issues that must throw the entire case into doubt. As Professor Myles Allen says:

> the use of the current GWP method to assess methane's impacts could mean eliminating practices, such as ruminant agriculture, that are not actually causing global warming.

Professor Myles Allen has argued for an improved methodology called GWP*. Using this shows that methane emissions from ruminant livestock should not be our focus. For example, a 2021 study using GWP* found that the entire European sheep and goat herd was climate neutral, with zero net emissions between 1990 and 2018, even before accounting for carbon sequestration. When we take account of the carbon sequestration provided by these pasture-reared livestock, the European sheep and goats are reducing atmospheric carbon and its warming impact. Professor Ian Scoones comments, 'This has important implications for mitigation. While CO_2 emissions [from fossil fuels] must reach net zero to stop temperatures increasing, methane emissions [from ruminant livestock] can be sustained indefinitely.'

So why are some people still arguing that we should reduce ruminant livestock numbers? Recent IPCC reports suggest that the world is going to exceed the 1.5 degrees temperature rise flagged by the 2015 Paris Agreement. Trying to avoid this by reducing fossil fuels is like trying to turn a supertanker at sea, because CO_2 remains in the atmosphere so long. The effect of cutting methane emissions is quicker. So, although ruminants themselves do not increase atmospheric greenhouse gas levels, cutting ruminants could theoretically provide a quick way to offset the fossil fuel emissions. But this would provide a one-off, short-term benefit.

And have you ever wondered how climate scientists measure how much methane cows and sheep burp out when they are roaming the hills? Quite tricky! To make life easier for the scientists the measurements are actually taken from individual animals kept in glass tanks in laboratories. As Ian Scoones explains:

> The chambers provide a highly artificial environment with large well-bred animals receiving harvested agricultural forage and manufactured feeds. These are very poor models for the metabolism and emissions of extensively grazed native livestock in Africa, Asia, South America, or even for extensive beef and sheep farming in the uplands of Britain. Empirical studies on local animals have shown much lower emissions than the standard models, because the animals are generally smaller, physiologically adapted to their environment and able to selectively graze from a diverse variety of pasture plants.

If ruminant livestock are not the cause of rising atmospheric methane levels, what is? Climate scientists are considering several possibilities, including increases in paddy rice hectarage, biomass burning, deforestation and wildfires. And we seem to have hugely underestimated the volumes of fugitive emissions leaking from oil, gas, fracking and coal mining infrastructure, and petrochemical processing. A 2016 study found that methane emissions from fossil fuels were 60% higher than previously thought. There are also worries about increasing methane emissions caused by the degradation of natural wetlands, often caused by draining them to grow crops.

One of the greatest risks of methane emissions is the melting of the arctic permafrost, below which are huge natural reserves of methane that are already escaping as the permafrost is melting. This is another example of where old answers may be the best. During the Pleistocene period (2.6 million years ago until the end of the last glacial period 11,700 years ago), the arctic tundra was covered in steppe grassland, created by the presence of large grazing animals such as mammoths and including ruminants such as bison. Though we can't replace the mammoths, new restoration projects such as *Pleistocene Park* are reintroducing grazing animals, including bison, reindeer and horses. The grazing animals remove tree saplings and woody vegetation, preventing the steppe grassland from turning to scrub and forest. By maintaining this steppe landscape the grazing animals encourage winter snow and ice cover, which would otherwise be eroded by scrub and forest growth. The snow and ice increase the *albedo* (the reflectivity) of the landscape, so that more solar radiation bounces back into space (woody vegetation and trees reduce the albedo). It is hoped that the reintroduction of more grazing animals will help reduce permafrost melting and methane leakage. And this is not necessarily about rewilding. Meat eaters, including wolves and humans, have long been part of these landscapes and today, pastoralists, such as reindeer herders, can still play a valuable role.

FEED EMISSIONS

According to FAO assessments the biggest proportion (55%) of emissions from beef cattle farming comes from growing feeds such as grains. These include emissions from fuel used during harvesting and agrochemical manufacture, and emissions from feed crop wastes. But 90% of beef cattle worldwide are reared at pasture, with little or no supplementary grain feeding. And ruminants can be reared completely without grains. It is crop growing that produces the harmful emissions, whether we grow those crops for livestock or human consumption.

The assessments also omit to include the other side of this balance sheet: all over the world, cattle and other livestock consume huge quantities of crop residues that people don't eat, such as straw, rice hulls and vegetable wastes, thereby mitigating large volumes of greenhouse gas emissions from decaying crop waste.

Every autumn, when I worked in the fenland crop-growing areas of eastern England, I would see tens of thousands of sheep being brought in from the distant hills of Leicestershire and Derbyshire. Brassicas produce a lot of unmarketable waste, like stalks and outside leaves that are left on the fields after harvest. The sheep clear up thousands of acres of this every year. Tonnes of crooked carrots and misshapen potatoes are also fed to cattle. Without livestock to eat these crop residues, they would rot and produce emissions in the fields, or have to be ploughed in using lots of diesel with more emissions. And a bonus is the dung that the animals deposit on the fields, increasing the soil carbon stores and reducing the fertilizer requirement of the next crop.

The FAO also counts nitrous oxide (N_2O) and methane emissions from fertilizers and manures that are applied to pasture and feed crops. But most of these emissions come from slurry (liquid manure) storage and application. While slurry is commonly produced from intensive dairy and pig farming, most beef and sheep farming in Britain and elsewhere produces solid manure, which has lower emissions. And again, the FAO assessments fail to account for the carbon sequestration benefits of animal dung being incorporated into the soil.

Also included in FAO estimates of livestock feed emissions are those from the deforestation of rainforests owing to pasture expansion in the Amazon. We share the public concern about deforestation in the Amazon and elsewhere. We should not cut down the Amazon to rear beef, any more than we should cut it down to plant soya, cotton or oil palm, as we are. But beef rearing elsewhere, such as in Britain, does not cause deforestation. While some historic deforestation may have occurred to provide grazing for livestock, a lot of it was to provide fuel, building materials and land for arable cropping.

A FAT FLAW IN THE FAO FIGURES

The way the FAO assessments compare beef emissions with those from other livestock, like chicken and pork, is deeply flawed. They compare emissions per kilogram of protein produced. This leads to the conclusion that beef emissions are the highest, and chicken and pork thus appear much more emissions-efficient. This is not surprising since chicken and pork carcases are much leaner than beef. The comparison gives a distorted picture because it only values the protein content of the food and ignores other nutritional components. Compared with chicken and pork, beef is much richer in micronutrients such as haem iron and B vitamins, but the most misleading aspect of the comparison is that it completely ignores the nutritional value of fat. A beef or lamb carcass has a higher fat content compared to chickens and modern pigs. This fat is a rich and healthy source of energy and essential micronutrients. It makes a valuable contribution to the diets of people all over the world. So, while the FAO comparison makes chicken emissions look much better than beef, chicken is a poorer-quality food for humans.

Having said that, it is still worth remembering that all these meats are nutritionally superior to plant foods. Unlike vegetable fats, animal fats are a well-balanced source of essential fatty acids and fat-soluble vitamins. And animal-sourced foods provide essential vitamins that plants cannot provide, such as Vitamins B12 and K2. Plus, animal protein is nutritionally superior to plant protein.

Tragically, the high nutritional value of animal fat is poorly recognized in Western countries; a wasteful hangover of the diet–heart hypothesis that we discussed earlier in this book, which has so negatively influenced our dietary guidelines. On average, 28kg of fat is trimmed and discarded from every British beef carcass, not because there is anything wrong with it, but because consumers have been misadvised to avoid eating it. It is time that governments and consumers recognized the value of eating animal fat.

People need to eat. If we eliminate livestock farming and eat only plant foods, this will reduce the overall amount of food produced, because so many ruminant animals are reared on land that can't grow crops. And because crop foods are less nutrient-dense, crop-farming emissions would increase without producing equivalent human nutrition. It is simply impossible to equate and compare animal foods with plant foods. So, the question is not whether we should eliminate livestock farming, but how should we rear livestock to minimize net emissions and other negative impacts? And at the same time, how do we reduce emissions and other harms from crop farming? One of the most obvious ways is to put pasture-reared livestock back into crop rotations, to restore carbon to the soil and reduce crop agrochemical use.

Sadly, this false belief about animal fat has also changed the landscape of farming. During my lifetime I have witnessed the almost complete replacement of British native cattle breeds with bigger, faster-growing and leaner European breeds. When I started work as a farming adviser in the 1980s, the Hereford was still the most commonly used beef bull, and many farms kept Devon cattle, Lincoln Reds and other natives. The great taste and tenderness of their meat came from their slower growth rates and higher fat content. But by the end of the 1980s most commercial beef herds had started replacing these breeds with the leaner French Charolais and Limousin, and the Belgian Blue.

Today, most of our native cattle are rare breeds, kept by a few die-hard enthusiasts. Ironically, considering the environmental accusations against cattle and sheep, they are often kept in our most environmentally precious and protected landscapes, because they support and enhance wildlife. Our native breeds are especially good in these situations, because they are hardier than the non-native breeds and do well without the need for grain feeding.

This massive change was partly because of the higher growth rates of the European breeds (although usually requiring more grain feeding), but the other major driver was the lies we've been told since the 1970s about animal fat and heart disease. As a result of this damaging public health messaging supermarket buyers began demanding less fat. Even today supermarkets trim off fat and promote 'lean mince'. And the traffic light food labels still give us red warnings about perfectly healthy animal fat. Abattoirs in turn began trimming fat from carcases, and farmers were penalized for what were classed as 'overfat' cattle. Since beef and sheep farming has long been the least profitable of farming enterprises, farmers responded to these market forces because they had no choice economically. Even the Aberdeen Angus, with its global reputation for quality and taste, could not escape *improvement* by being crossed with breeds like the Limousin to make it leaner.

INTENSIVE VERSUS EXTENSIVE LIVESTOCK FARMING

One of the most extraordinary recommendations of the FAO global assessments was that changing from extensively reared red meat (pasture-reared cattle and sheep) to intensively reared white meat (housed, grain-fed pigs and poultry) would reduce emissions and climate change.

The rationale behind this was that intensively reared livestock grow quicker. A broiler chicken is ready for slaughter at six weeks old, and a barley-beef bull at a year old. But a pasture-reared beef cow can take two to three years to reach slaughter weight. In Africa and Asia, native cows and sheep often take even longer to fatten. The FAO assessments claim that the longer growing period produces more emissions, because the animal produces emissions just in maintaining itself each day, so taking longer to fatten increases the maintenance emissions.

This recommendation ignores the fact that intensive livestock like pigs and poultry depend on grains from arable land that could otherwise be used to grow crops for direct human consumption, whereas pasture-reared cows and sheep can be reared on marginal land that can't grow crops. Intensive livestock also produce much higher fossil fuel emissions, especially CO_2, because of their reliance on grains, housing and mechanization. Whereas pasture-reared livestock farming has very low fossil fuel usage, and the main emission is biogenic methane that plays no part in global warming. Slurry (liquid manure) produced in intensive livestock farming also causes more water and air pollution.

Implementing this recommendation to intensify livestock farming would be disastrous for the environment and for human societies around the world. It would most especially harm low-income grassland farmers and the world's 1.3 billion pastoral herders, already some of the most marginalized people in the world. Replacing grazing livestock with intensive livestock would wreck cultures and livelihoods and inevitably result in millions of people losing access to their lands. From an emissions viewpoint it would be completely counterproductive, replacing harmless biogenic methane with high-impact fossil fuel emissions, and forgoing the benefits of carbon sequestration.

What we eat and how we farm matters. But it's the how, not the cow! Intensive farming systems are environmentally and ethically unsustainable whether they produce livestock products or plant foods. But ruminant livestock such as cows and sheep can be reared at pasture, with low inputs and low net emissions. The role of pasture-reared ruminants in carbon sequestration offers us our greatest opportunity to produce healthy food while reversing climate change.

As we have seen, soil and grasslands play key roles in the carbon cycle. We'll dig deeper into the soil next: what is soil? What does it do? Why is it so important for us to understand it?

CHAPTER 19

THE ANSWER LIES IN THE SOIL*

How cows reverse climate change ... and crops cause it.

'The remaining soil is so fine I am able to sink my hand into it right up to my forearm. I bring up a handful, open my fist and watch it blow away in the wind across the huge flat field. This soil has almost nothing left in it. Everything the crop will need must be added.'

Sarah Langford, *Rooted*

AS A NEWLY FLEDGED FARM ADVISER, I spent many summer days striding across cattle and sheep fields, a little blue ruling stick in hand – a *sward-stick* (sward being an area of grass-covered soil). Every few paces I'd stoop down with the swardstick and measure the grass height. Nearby a red cow would lift her head quizzically mid-munch, as if to say, 'And you humans think animals are dumb!' Then, looking around, I'd see the farmer following me with the same bemused expression! Earlier that spring, I had handed out the little blue sticks to farmers across the county, along with glossy grassland calendars. I explained our latest research, and how to use these not-so high-tech tools to manage the grazing height and know when to move livestock on and off different pastures through the grazing season. Who'd have known that grassland science was such a charming art.

This sward-height management system worked brilliantly. The grass sward grew dense, never overgrazed or undergrazed. The animals grew fast and milked well, farmers made more silage and hay than usual, and we regenerated

* Arthur Fallowfield, *Beyond Our Ken*.

thinning areas of pasture, avoiding soil erosion and the need for reseeding. And though neither the farmers nor I knew it at the time, we were doing far more than rearing meat; we were working with nature's age-old dance between soil and atmosphere, grass and grazing animals, stowing carbon safely below ground, unknowingly helping to regulate the climate.

UN climate bodies like the IPCC focus almost entirely on emissions, but Earth's climate regulation is not just about emissions. Natural cycles such as the carbon cycle and the water cycle are vital too. As we shall see here, the world's vast grasslands, grassland soils and herds of grazing animals form a system that is especially important in supporting these natural cycles and the balance of the Earth's climate. Yet we are in danger of disrupting this system and unwittingly accelerating the environmental crisis we claim to be trying to prevent.

In chapter 18 we looked at the biogenic carbon cycle and saw how there is a continuous flux of carbon from the atmosphere, through plants and grazing animals and back into the atmosphere. But not all of this carbon goes back to the atmosphere. Sequestration pathways divert some carbon to be stored in the soil. So, the amount of carbon that is held as carbon dioxide (CO_2) and methane (CH4) in the atmosphere depends not just on emissions, but also on how much carbon is sequestered into the soil.

Despite its recent bad press, carbon is not a bad thing, and nor is CO_2. All life on Earth contains carbon and plant growth depends on CO_2. What matters is how much carbon is held where. Too much in the atmosphere as CO_2 and CH4 may contribute to warming. Too much CO_2 in the ocean causes harmful acidification. But carbon that is stored in the soil, as organic matter, helps to regulate the climate. It also makes the soil fertile and supports plant growth and the life of a vast ecosystem of tiny soil creatures; a quarter of Earth's living organisms live in the soil. So, carbon sequestration and storage in the soil is essential to all life on Earth.

Earth's soil holds 2,500 gigatons of carbon, which makes it the planet's second largest carbon store, after the oceans. This is five times more carbon than is held in all Earth's forests, and more than in the planet's entire biomass and atmosphere put together.

The two most important soils for carbon storage are the peat soils in wetland ecosystems (which cover a relatively small area of the Earth), and the soils beneath the world's vast areas of grasslands. These two types of soil ecosystems are especially significant for two reasons: since they hold so much carbon, the degradation of these soils causes significant CO_2 emissions; on the other hand, good management of these two ecosystems offers us the greatest potential for carbon storage to reverse emissions through carbon sequestration.

UN bodies agree that carbon sequestration is a real thing, but as they have not agreed on a methodology to measure it, they generally ignore it. This focus on emissions while ignoring the biogenic carbon cycle and water cycle is leading us in completely counterproductive directions.

Farming activities can either increase or deplete the amount of carbon that is held in the soil, and therefore how much ends up in the atmosphere. Since soil holds so much carbon, soil mismanagement can release large amounts of CO_2 into the atmosphere. It also threatens global food security through soil erosion and desertification. We have already lost 15–20% (500 gigatons) of carbon from the Earth's soil since humans began farming 10,000 years ago. A third of the carbon added to the atmosphere in the last two centuries has come from soil degradation. Though inappropriate grazing management can be a factor, the activities that cause most soil damage and CO_2 emissions are crop farming and the conversion of grasslands, forests and wetlands to crop farming.

ARABLE DEPLETES

Arable soils tend to contain significantly less carbon compared to grassland soils, because arable farming depletes soil organic matter. Up to 40% of soil carbon is lost within two years of ploughing grassland and converting it to arable cropping. Conversion of grassland to arable farming is a major contributor to CO_2 emissions, making up 25% of agricultural emissions in the UK, for example. But when land is converted in the other direction, from arable to grass, carbon is returned to the soil.

Most arable crops are annuals that are sown every year. The seedbed is prepared each year, usually by ploughing or other methods of tillage. This tillage disturbs the soil and exposes its organic matter to the air every year, and this results in much of its carbon content being converted to CO_2 and emitted into the atmosphere. In addition to this, arable land is often left bare for much of the year, and crop plants are widely spaced. This exposes a lot of the soil to the air for a lot of the time. The fungal network, earthworms and other invertebrates that support carbon sequestration in soil are also repeatedly damaged by crop tillage and agrochemicals (only a tiny proportion of arable land is farmed without agrochemicals). Every application of fertilizer or pesticide involves heavy machinery travelling over the land, compacting the soil, reducing aeration and drainage. All this gradually kills the soil ecosystem and turns arable soils into CO_2 emitters rather than carbon stores like grassland soils.

Exposed and degrading arable soils, England, May 2024.

Perhaps the most infamous example of soil damage and desertification was the Dustbowl disaster of the 1930s. Millions of acres of North American farmland turned to dust and blew away, after the prairie grasslands that had once been sustained by 60 million bison were ploughed to grow arable crops. This man-made catastrophe was immortalized in John Steinbeck's novel *The Grapes of Wrath*. The Dustbowl led to the displacement of 3.5 million Americans between 1930 and 1940. A third of the topsoil in the American Midwest is thought to have been lost during the twentieth century as a result of the conversion of grasslands to crops.

Desertification occurs when soil loses its organic matter. The soil dries out and becomes vulnerable to erosion by wind or rain. In temperate countries few of us think about desertification. Nevertheless, the arable soils of East Anglia have already lost three-quarters of their topsoil, and 99 million acres are at risk of desertification in the EU, along with 74% of North American land. In Australia 1.2 million acres are at risk, and 2.4 million acres in Africa. Arable crop growing and inappropriate grazing management are both contributing.

Peat soils, as found in lowland wetlands and boggy upland moorlands, are extremely important for climate regulation (and biodiversity) because they comprise more than 75% organic matter. The Earth's peatlands are its most important carbon store. Under natural conditions peat remains waterlogged and covered by vegetation. But nearly 90% of the world's wetlands, and their peat, have been lost in the last 300 years, mainly owing to land drainage and tillage for crop farming, or peat harvesting for horticultural compost. Both practices expose the peat to air so that its carbon is rapidly released as CO_2, and the peat shrinks.

England has 1.4 million hectares of peatland remaining, and these hold as much carbon as the same area of tropical rainforest. The fens of East Anglia are the most important peatlands, but arable farming has reduced the amount of peat by 84% since 1850. The peat surface is dropping by 2cm per year, and producing 400,000 tonnes of CO_2 emissions, equivalent to 1% of annual UK aviation emissions. Globally, carbon emissions from peat drainage and burning increased by 35% between 2000 and 2018.

Most farmers are very aware of soil damage risks and do their best to reduce them. Many arable farmers have introduced cover crops to protect the land from erosion when it would otherwise be bare. Others have given up the plough and invested in new and expensive direct seed drills and minimum tillage (min-till) machinery that reduce soil disturbance.

The UN has tried to promote min-till as a way of increasing soil carbon sequestration on arable land. But while min-till can reduce soil erosion, it does not enhance soil carbon sequestration. It has even been found to reduce soil carbon levels, and crop yields in some cases. Britain's world-renowned Rothamsted Research organization carried out a review of min-till studies, including the '4 per 1,000' initiative promoted by the UN with the aim of increasing soil organic matter by 4% each millennium. Rothamsted found: 'the 4% annual rate of SOC (Soil Organic Carbon) increase is unachievable in most agricultural situations and cannot be regarded as a major contributor to climate change mitigation'. However, Rothamsted found that soil carbon sequestration could be increased in other ways. One way was to plant woodland in place of the arable crops; the other was to replace arable crops with pasture and grazing livestock for most of the rotation.

Green mulches and green manures are another tool used by smallholders to restore soil organic matter. Some advocates of plant-based diets even believe that these make livestock completely unnecessary in crop farming. These crops are cut and left to mulch on the soil surface or ploughed back into the soil before the next crop. These techniques were once standard parts of traditional 'mixed-ley farming systems' long before industrial agriculture or veganism. The crucial difference, however, was that cattle and sheep were always included, grazing the green manure crop before it was ploughed in. Green manures and mulches may slow the soil depletion but cannot prevent it without animal manure. Eventually, the crops have to be moved – literally to pastures new! And green mulching does not work well in drier parts of the world with seasonal rainfall, because it doesn't rot down quickly enough unless ploughed in.

Organic and regenerative farmers know that the only way to replenish and restore arable soils is with pasture, livestock and manure. But a plant-based world has no livestock. If people wish to continue eating plant foods, and to do so in the least environmentally harmful way, then we need grass in our crop rotations, and ruminant livestock to graze it, recycle the crop by-products, and provide manure to feed the crops and put carbon back in the soil.

GRASSLAND RESTORES

Fortunately, when it comes to carbon, what goes up can also come down, through the process of sequestration. World-renowned soil scientist Dr Rattan Lal estimates that sequestration could restore two thirds of the carbon that has been lost from world soils, and that this could offset 15% of global fossil fuel emissions.

Grassland soils are especially important for carbon sequestration. They sequester even more carbon than temperate forests. And while forests store most of their carbon above ground in the tree biomass, where it is vulnerable to fire, grasslands store it safely below ground, where it does not release CO_2 provided the grassland is not ploughed up. There has been a bit of a tree-planting craze in recent years. Every corporate entity has been doing it to prove their sustainability credentials. But ripping up grassland to plant trees is counter-productive for climate change mitigation. Converting arable land to forest and grassland would make far more sense. In some areas a combination of trees and pasture – known as silvopasture or wood pasture – can work well. The trees, grassland and livestock all contribute to carbon sequestration, and the land can provide meat and milk at the same time as fruit and nuts, or timber. This is the ancient practice that helped form historic landscapes such as England's New Forest.

Permanent grassland, with well-managed grazing animals, provides high rates of carbon sequestration. Most beef cattle and sheep are reared on permanent grassland that is rarely if ever ploughed, and few if any agrochemicals (artificial fertilizers and pesticides) are applied. Such grassland farming provides the most complete and continuous ground cover, so that the soil is rarely exposed to air and CO_2 losses are minimal. The deep, undisturbed roots of permanent pasture enhance soil fungal activity and soil structure and add carbon to the soil. And grazing livestock replenish the soil with their dung – dung that is pre-inoculated with microbes from the ruminant gut. This helps make the nutrients available to soil organisms. By contrast, when vegetation dies without being eaten by grazing animals, it takes much longer to be incorporated into the soil and a large amount of its carbon content is released to the air as CO_2.

REGENERATIVE GRAZING

Well, you've already heard about my younger days with a swardstick, providing entertainment for the local livestock and their farmers. But those ever-sunny days of memory were a little while ago.

Today there are new ideas and debates on regenerative grazing systems. Making sure that animals move before they graze plants a second time is key – an art that the world's pastoral herder groups have long practised. If animals are allowed to stay in one place too long, they can keep re-grazing the same plants too low, nibbling off the succulent new shoots each time the plant tries to regrow. This hampers the plant's ability to make carbohydrates by photosynthesis, forcing it to draw on energy reserves stored as starch in its roots. Gradually, plants weaken and die. Patches of bare soil appear. Soil organic matter begins to vaporize, transferring carbon into the air, along with the moisture the soil once held. The dry soil begins to blow away and the pasture gradually turns to desert. This is what I believe has happened in Mongolia.

Regenerative livestock grazing has shown its potential, not just to prevent desertification and soil carbon loss, but to reverse them. It also increases biodiversity and restores ecosystems.

Award-winning Zimbabwean ecologist Allan Savory has been one of the pioneers of regenerative grazing. Early in his career, Savory accepted the common belief that soil erosion and desertification were caused by having too many animals. But then he saw that there was no improvement even when thousands of elephants and other herbivores were culled. By observing wildlife, Savory realized the critical factors were movement and grazing pattern rather than numbers. He developed a system called holistic planned grazing, which has been used to restore soils and reverse desertification all over the world. Livestock are 'mob-grazed' as a dense herd that moves frequently to new pasture, leaving the grazed area to rest. The grazing pressure and trampling action of the animals' hooves, breaks the soil surface, pressing plant material, dung and urine into it. This stimulates soil organisms, incorporates organic matter into the soil and increases water retention. The quick movement and long rest periods encourage strong, deep roots and avoid depleting the plants' reserves.

Scores of studies have found that regenerative grazing, and converting croplands to grazed grasslands, sequesters large amounts of carbon into the soil, offsetting agricultural emissions, restoring soil that has been damaged by crop farming and reversing desertification. A global transition to organic (or regenerative) farming, with grass and livestock in crop rotations, would increase carbon sequestration by 1.5 million tonnes per year over the next twenty years, mitigating 11% of all man-made emissions, according to the UK's Soil Association.

So, increasing the global acreage of regeneratively grazed grasslands and putting grass and livestock back into crop rotations would be a far better strategy for reducing agriculture's carbon footprint than cutting ruminant numbers to reduce methane.

TO SHARE OR TO SPARE?

Some environmental activists advocate that we should follow a land-sparing and rewilding strategy, removing all forms of agriculture from some land and setting it aside entirely for nature. The flipside, and downside, of this is that agriculture would have to be intensified on other land, increasing the environmental harms in these areas. Ultimately, this land-sparing philosophy drives some advocates to conclude that all agriculture is harmful, and people should give up plant foods as well as meat, and be made to eat synthetic, factory-produced nutrients. But this opens a whole new Pandora's box of problems, such as how to meet the energy demands of a food production system that has been de-linked from the free solar energy that crops and grazing livestock harness.

A more humane and sustainable approach is that of land-sharing: low-impact farming methods such as regenerative and organic farming that sustain the soil and integrate food production in ways that sustain wildlife and the natural environment. It means recognizing the immense value of grassland farmers and pastoral herding communities, from the west of Britain to America and Mongolia, and rearing meat and milk to feed the world, from land where crops cannot. Furthermore, it means growing the crops we want more sustainably, by returning to organic crop rotations that include grass and livestock.

KEEPING THE WATER CYCLING

So far, we've talked mostly about how soil, grasslands and livestock support climate regulation through a healthy carbon cycle. But their role in the water cycle is interconnected. When water evaporates it cools the surroundings, and when it condenses as rainfall it warms them. So, the water cycle helps regulate temperatures locally and globally.

Dung from grazing livestock supports local water cycles by increasing soil organic matter and supporting all the creatures in the soil ecosystem, which in turn produces good soil structure and permeability to improve rainwater percolation. This increases the land's water-holding capacity, supports vegetation growth, prevents drought and recharges local groundwater supplies. As a result, rainfall is held in the soil and vegetation, and gradually released back to the atmosphere by plants (evapotranspiration), rather than rapidly running off into rivers. This evapotranspiration from trees, wetlands and grasslands has a cooling effect. It also causes low-level clouds to form, which shade the land from the sun's radiation. Arable land does the opposite of this, reducing the amount of water held locally, disrupting local cloud formation and water cycles.

Healthy soil is also critical for natural flood management. It acts like a sponge, absorbing huge amounts of water during heavy rainfall. Less water runs off the land to cause flooding. In dry periods, it reduces drought by holding more water in the soil for longer. It also improves water quality by reducing soil erosion that contaminates watercourses. So again, cattle and sheep grazing can help prevent flood risk and water contamination, whereas the damaging impacts of crop growing on soil contributes to these problems.

The world's grasslands provide a wide range of ecological benefits such as flood mitigation, soil erosion control, carbon storage, climate regulation, biodiversity and pollination. Cattle and other grazing livestock are an integral part of the grassland ecosystem and its benefits. These invaluable grassland ecosystems have declined worldwide, mainly because of their conversion to arable land. One study concluded:

> 'Covering twice the area suitable for crop production, the potential role of grasslands for global food security should be better recognized. Ruminant grazers are efficient converters of non-edible plant biomass such as grass into human-edible energy and nutrients and can do so without competing with humans for use of crops for foodstuffs.'

HOW COWS CARE FOR HABITATS AND CARBON

Cattle and sheep contribute to the sustainability of many of the world's most beautiful landscapes and habitats. Though fields of wheat and cabbages may have their charms, the green hills and uplands of Britain and range grasslands around the world are some of the most attractive places to visit, and some of the most important for wildlife. Cattle, sheep, goats and other livestock sustain these rural communities, and their grazing contributes to land management in ways that enhance both the landscape and ecology. The restoration of Woodchester Park, in Gloucestershire, is one example. The park is home to rare horseshoe bats. Its valley has been restored to meadow and cattle have been reintroduced, as their grazing creates a mosaic of grass heights, which encourages a diversity of wildflowers and insects. Very importantly, cow pats provide the perfect food for dung beetles, which are the main food source for horseshoe bats.

Britain's upland grasslands are of international importance, and they are home to 15% of the world's blanket bog, with deep peat soils that store large amounts of carbon. The varied wet and dry habitat and different vegetation types of moorland provide habitat for many significant species, including a quarter of the world's curlew population, as well as many rare insects. Despite the poor management of Dartmoor described earlier, carefully managed grazing with cattle and sheep is invaluable in these environments. The grazing maintains a mosaic of habitat and species diversity. And by preventing too much shrub growth, which would dry out the bog, grazing at the right time with the right number of animals provides good growth conditions for the sphagnum moss which builds the peat. So again, cattle and sheep help to store carbon and combat climate change. In Ireland too, well-managed livestock grazing plays an important role in peat moorland management, the prevention of wildfires and conservation of moorland birds like the hen harrier.

In California, wildfires torched 1.7 million hectares of land in 2020, destroying trees and homes and releasing huge volumes of greenhouse gas emissions. Since then, flocks of sheep and goats have been brought in. They prevent dry vegetation from building up, and so reduce the fire risk and emissions. The animals avoid the need for expensive mechanical vegetation clearing, or harmful herbicides and the associated emissions from fossil fuels. On top of that, their grazing encourages biodiversity and replenishes soil organic matter and moisture.

It's time that we recognized how much crop farming is damaging our soil and contributing to emissions. More crops and plant-based diets are not the way forward. It is grasslands, grassland soils, and pasture-reared livestock that offer us our best opportunity to reverse climate change.

CHAPTER 20

WHEAT IS MURDER

Death and destruction in plant agriculture

'How could intelligent beings seek to control a few unwanted species by a method that contaminated the whole environment and brought the threat of disease and death even to their own kind?'

Rachel Carson, *Silent Spring*

THE DESIRE NOT TO CAUSE SUFFERING IS one of the concerns that commonly motivates people to give up meat. We not only respect this wish to do no harm, we share it. Anti-meat campaigns often frame the issue thus: eating meat involves animal suffering and death, whereas eating plants doesn't. But are we so sure about this? The truth is that there are lots of reasons why we kill animals; eating them is just one. It makes no sense to believe that killing animals for their meat is wrong, but that killing animals so that we can eat vegetables or grains is all right. But in our modern society, so distanced from the origins of our food and the natural environment, it is easy and convenient to pretend that sustaining life does not involve death.

In this chapter we look at the other side of the coin, that plant-based diet advocates conveniently ignore: the death, distress and ecological damage that plant-based diets cause, and why pasture-reared beef and lamb are the most ethical choices of all foods.

The questions involved in the decision to eat meat, or plants, or both, are much wider than anti-meat campaigns would have us believe. The question *did this meat come from killing an animal?* aims to deter us from understanding the full picture of harms and benefits involved in crop and livestock farming. The question *did this bread come from killing animals?* must also be asked. In fact, as we shall see in this chapter, crop production causes more animal death and suffering than pasture-based livestock farming.

The invention of the plough began four millennia of soil degradation and ecosystem destruction, but the world has never before seen ecological destruction of the scale caused by crop farming during the last hundred years. Few of today's generation realize how arable farming has changed during the last century, or how dramatically this has reshaped our countryside and set in train a cascade of environmental consequences.

Arguably, the most damaging and unsustainable change has been the removal of livestock from arable rotations. Until the mid-twentieth century, arable crops were typically grown in rotations that included pasture and grazing livestock. Two or three years of crops were followed by two or three years of pasture with cattle or sheep grazing.

Arable crops degrade soil, release its stored carbon to the atmosphere as CO_2 and drain it of fertility. When crops are grown year after year, the soil degradation accumulates. Crop diseases also build up and depress yields. Historically, grass and livestock were incorporated in arable rotations, breaking the cycle of crop diseases, replenishing the soil carbon stores, and restoring soil fertility.

Tragically, during the last century, all over the industrialized world, pasture and grazing livestock have been removed from crop rotations. With the coming of artificial fertilizers and pesticides, and bigger, better machinery, it seemed that animals and natural manure were unnecessary complications; farmers simplified their farming systems and increased profits by changing to crop rotations without grass and livestock. But the consequences for our soils and ecosystems were unforeseen and are still largely unappreciated.

The removal of livestock also meant bigger fields were possible. This in turn meant fewer field boundaries were needed. Across Britain and Europe thousands of miles of wildlife-rich hedgerows were removed and water-filled dykes (traditionally used as field boundaries in fenlands) were filled in, to create larger fields for intensive crop production. Today, crop-growing areas, like the east of England, are industrial landscapes, not of chimneys and smoke, but of wheat, sugar beet, cabbages and potatoes, all swathed in a mist of pesticides.

When I began working as a crop agronomist in Lincolnshire, I learned early on that crop production involved killing a lot of animals. Most of my time was spent trudging across fields, identifying insects, fungal diseases and weeds – then advising farmers which of the mind-boggling range of insecticides, molluscicides, fungicides and herbicides would kill these most cost-effectively.

It was an education in ecological devastation. The unnoticed collateral damage of this crop production has been the bees, butterflies and earthworms, and then the mammals and birds, and entire ecosystems that depend on the small lives at the base of the food chain. The soil damage and habitat loss have added to the uncounted cost of our daily bread and five-a-day. All to supply a disconnected nation's demand for cheap food. The only respite for wildlife was around the fringes of this industrial cropland, where the land would not allow crop production; here there remained oases of grassland, where beef cows and sheep grazed unobtrusively alongside wildlife, with no pesticides and no ploughs.

Some people claim that they always buy organic fruit and vegetables. Yet the Soil Association was founded in the UK in 1946, and still only 1–2% of UK cropland is managed organically. Some 98% of UK cropland is managed intensively with heavy and increasing use of agrochemicals. The figures are similar in the EU and USA. This is not down to uninterested or unconcerned farmers. It is market economics and the unwillingness or inability of consumers to pay the price, while often lamenting the harms of intensive farming. In practice, 80% of the registered organic land in the UK is grassland from which beef, lamb and milk are produced. Even if the grassland is not registered as organic, most beef and sheep are reared with little or no agrochemical use. Few consumers today appreciate that livestock are an essential part of organic (and regenerative) crop farming systems. Or that the most destructive farming systems are characterized by an absence of livestock.

ANIMAL DEATHS IN CROP PRODUCTION

All over the world, billions of birds, mammals, reptiles, amphibians and invertebrates are killed during crop growing. Those killed mechanically during tillage and harvesting are just the tip of the iceberg. Most are deliberately *controlled* – a euphemism for killing with guns, pesticides and poisons – because they are considered to be crop pests or vermin. Many more are lost owing to indirect pesticide impacts and habitat destruction. Thousands of species have been brought to the edge of extinction by crop production. Non-meat eaters generally ignore these deaths, as the animals killed for their bread don't usually appear in the sandwich.

By contrast, pasture rearing of beef cattle and sheep rarely involves tillage of the soil, because they are mostly reared on permanent grassland. Nor does it involve pesticides or vermin control. As a result, it enhances wildlife habitat and biodiversity, as we saw in chapter 19.

Perhaps we think the animals that die for our bread and veg don't matter as much as cows and sheep, but why is a cow's life more important than a wild animal's life? Even if that wild animal is *just* a rodent or an insect, it is a life, and it contributes as part of a food chain that other animals rely on.

Some people take the view that if they don't personally kill and eat animals, they are not responsible for any animal deaths. This is a form of magical thinking or wilful ignorance. In practice, eating plants means killing animals just as eating meat does. But whereas farm animals in the UK and most developed countries are killed humanely, subject to regulations and with a vet in attendance, arable farming inhumanely destroys vast numbers of animals that are considered a threat to crops.

In the Mediterranean, for example, millions of songbirds get 'vacuumed' and killed by intensive harvesting machines that are used to strip olive trees of fruit every year. Many farmland birds, rodents and other mammals are also killed during grain harvesting all over the world, although, as we shall see below, this is the least harmful aspect of crop production.

In his book *On Eating Meat*, Australian chef and farmer Matthew Evans describes how growing 75 hectares (400 tonnes) of peas involved the killing of 150 deer, 800–1,000 possums, 500 wallabies and countless rodents each year. And a small orchard required 100 or more possums to be killed each year.

Twenty-five times more sentient animals are killed to produce a kilo of crop protein compared with a kilo of protein from beef. This is because meat is much more nutrient-dense than plant foods and a cow is a large animal, so one cow life produces a lot of human nutrition, whereas huge numbers of small animals must be cruelly killed (by shooting, trapping and poisoning) to produce the equivalent amount of plant nutrients.

Adult humans need to eat around 60g of protein per day, which equates to about 300g of beef. One cow provides enough protein to sustain a human for two and a half years, or about 930 people for a day. The number of insects, birds, rodents and other mammals that die to grow the equivalent amount of protein from crops is many times higher. But bear in mind that it doesn't matter how much plant food one eats or how many animals are killed to grow it – plant foods cannot provide all the vitamins that are needed for human health. On the other hand, beef fat, offal and bone marrow provide all the vitamins that humans require for good health; and they do so without the anti-nutrients and other substances in plants that cause ill-health.

It is difficult to keep track of the animal cost of crop production. While reliable data on animals slaughtered for meat is available in developed countries, crop farmers are not required to keep a body count. Similarly, we noted in chapter 18 that there are no comprehensive assessment methods for how crop production impacts climate change. Criticizing livestock farming is easier if you don't bother measuring the harms of crop production. Nevertheless, we will see that the scale of death and destruction is evident in the wildlife species losses and declining populations being recorded all over the world.

All wild birds are legally protected in Britain, yet millions are killed under licence to protect the crops they would otherwise eat; the very crops that replaced their natural habitat or the pastureland where they used to be safe. This includes around 3 million woodpigeons each year, and many more feral pigeons, collared doves and stock doves. Geese, ibis, gulls, crows, jays, rooks, jackdaws and magpies are some of the other birds that are slaughtered to protect British crops. The story is the same everywhere: in the Netherlands, half a million geese are killed to protect crops each year, while around 40,000 native ducks are destroyed annually to protect rice fields in New South Wales, Australia.

In addition to the deliberate killing of birds to prevent crop damage, many more are killed and harmed by the pesticide use and habitat destruction involved in crop farming. A large-scale study of house sparrows on Swiss farms found that all birds were contaminated with neonicotinoids, crop insecticides well known for their harm to bees. Even birds on organic farms that were not using the insecticide were contaminated with it, indicating that neonicotinoids are now ubiquitous in the environment, affecting many wildlife species and entire ecosystems. In 2018, neonicotinoids were banned across the EU because of their wildlife impacts, but the UK government introduced exemptions for sugar beet growers in 2023. Even when pesticides like DDT and neonicotinoids are banned, the agrochemical industry and crop farmers just replace them with different poisons, that is until it's learned that those too cause unintended harm.

Some birds are killed to protect other wildlife. Even the Royal Society for the Protection of Birds (RSPB) kills crows, magpies and the beautiful jay, because they predate other wild birds, such as the curlew, that we choose to protect. However, it was not the crows that first endangered the curlew, but human activity, especially the habitat destruction, disturbance and pesticide use involved in modern crop production. This is why curlew numbers have fallen most in lowland arable farming areas, and they are now thought of mainly as a bird of upland grasslands, where they can co-exist with cattle and sheep grazing.

What about those cuddly bunnies? British farmers and landowners are not only allowed to kill rabbits, they are legally required to do so, to prevent damage to neighbouring crops. British rabbit populations have sharply declined in recent decades. Myxomatosis is often blamed for this, but in fact the disease was exploited to reduce rabbit damage to crops. Although rabbits are not native to Britain, they have been here for at least a thousand years, and are so embedded in British ecosystems that their current decline is threatening the survival of rare plants, birds and other species that have come to depend on them.

Around 300,000 hares are also shot in Britain each year, for crop protection and sport (the sport often being justified by crop protection). Habitat destruction by crop farming is the other main cause of declining populations, whereas well-managed pasture provides good habitat for hares.

Billions of mice, rats, voles and other small mammals are killed by cropping operations, or vermin control. The bodies go uncounted, and the deaths rarely make *Guardian* headlines, or articles about plant-based diets. But the scale of the slaughter is apparent in the $2–2.5 billion annual global sales of rodenticides, and the plethora of companies offering vermin control services.

In Britain, deer have no natural predators, and the population is kept in check by an annual cull, with around 350,000 animals being shot. Wildlife organizations as well as farmers support the culling because the deer cause ecological damage to woodlands and woodland birds, like the nightingale, and around £10 million of damage to crops and commercial forestry each year. Some who advocate plant-based diets insist that it would be unethical to eat venison from culled deer, yet have no issue with consuming the bread or vegetables that those deer were slaughtered to protect.

There is little public criticism of the deer cull. But the culling of another British mammal, the badger, draws more public outcry than the killing of all the billions of birds, mammals, insects and other creatures that die for crop protection each year. Is it any coincidence that badgers are culled to protect cattle (from tuberculosis) and not, like rabbits and deer, to protect crops? The badger cull is a contentious subject, but whether it is effective and necessary or not, it is worth considering whether cattle or crops are the real threat to Brock.*

Badgers are a protected species, but they are not endangered in Britain. That's because they thrive in the pastoral landscapes of southwest England and Wales; landscapes created and maintained by cattle farming. The woodlands and hedgerows and the undisturbed pasture soils enriched by cattle dung provide perfect habitat and a bounty of the worms and other foods that Brock depends on. On the other hand, Brock is harder to find in England's eastern cropping areas, because of the loss of habitat and food sources as cropping has intensified and grazing livestock have been removed. So yes, let's find better

---

* Old English or Gaelic name for badger.

ways of controlling TB in badgers and cattle, by all means; but eliminating cattle farming won't help the badger.

Hedgehogs have seen precipitous population declines in rural Britain in recent decades, especially in the crop-growing areas of eastern England. Molluscicides and other crop agrochemicals are responsible, along with the destruction of hedgerows and other habitat. In southwest England, predation and competition from high badger numbers have contributed to Mrs Tiggy-Winkle's sad decline.

A COCKTAIL OF CARNAGE

In the rush to condemn cows and promote plant-based diets, many environmental activists today forget their origins. *Silent Spring* by Rachel Carson was published in 1962, and inspired the birth of the modern environmental movement. It documented how crop pesticides caused such devastation of wildlife populations that there would soon come a springtime with no buzz of bees and no birdsong. Carson realized that not only did these biocides kill target and non-target insects, but they also led to the collapse of entire ecosystems, by harming the birds and other wildlife populations all the way up the food chain. The book led to the banning of DDT, but sadly its full message has still not been understood. DDT and neonicotinoids have simply been replaced with other insecticides. Crop production practices, especially pesticide usage and habitat destruction, are still driving alarming declines in global wildlife populations.

In the USA, more honeybees die each year than all the livestock raised for slaughter and all the farmed fish because of crop-farming practices, especially pesticide use; 50 billion US honeybees were killed in 2019, more than six times the world's human population. California's almond-growing industry is especially implicated. Honeybees are taken to the almond groves by beekeepers to pollinate the almond blossom so that the trees produce nuts. 'It's like sending the bees to war. Many don't come back,' says Nate Donley, of the US Center for Biological Diversity. These honeybee deaths are just the tip of the iceberg, indicating the wider impacts on other insects and wildlife, as crop production undermines entire ecosystems. Most of the almonds sold in Britain are imported from California.

Besides the bugs that crop farmers target with pesticides, there are many inadvertent casualties whose loss we mourn – bees, butterflies and earthworms, to name but a few. But shouldn't we spare a thought for the target bugs that die too: the aphids, caterpillars, moths, wasps and slugs? Whether we regard the dead as beneficial or as pests, they all die in the same excruciating way. Most of the insecticides used in crop production are nerve and muscle toxins. The affected bee or wasp endures prolonged spasms of uncontrollable shaking, twitching and convulsions, slowly leading to paralysis and death. If we consider that these animals may be sentient, then this is a cruel way to die. And evidence is growing that many invertebrates (animals without backbones, including insects, spiders and molluscs) are sentient. In 2022 the UK parliament passed the Animal Welfare (Sentience) Act, based on a scientific review by a team at the London School of Economics, showing that invertebrates including octopi, squid, crabs and lobsters are sentient. Other studies indicate that bees, ants and even tiny fruit flies are sentient.

When one considers the vast numbers of these animals that crop pesticides kill every year, we should surely reconsider the ethics of eating crops grown with pesticides – which is almost all crops grown in the industrialized world. A report by the Wild Animal Initiative estimates that 3.5 quadrillion insects (3,500,000,000,000,000) are killed in the USA each year by insecticide use across its 100 million acres (40.5 million hectares) of cropland. That is around 438,000 times the human population of the entire world, or 100 million times the number of cattle that are slaughtered for meat in the USA each year. In the UK we have 10.9 million acres (4.4 million hectares) of cropped land. Only 1.65% of this is farmed organically, leaving over 98% that is farmed with agrochemicals. So, it is fair to estimate that UK crop production kills around 390 trillion (390,000,000,000,000) insects each year – 48,000 times the world's human population, or 150 million times the number of cattle that are slaughtered for meat in the UK annually.

So, is it any wonder that many of us have noticed that our car windscreens are no longer as splattered with bugs as they used to be? In 2017 a landmark German study was the first to confirm the hard reality of these observations. It found that flying insect numbers had declined by around 80% over a twenty-seven-year period. In the UK, a thirty-year study by world renowned Rothamsted Research found that insect populations had collapsed at all of four study sites. Other studies have identified the same trend globally. And they all point to the same causes: crop production, with its associated pesticides, fertilizers, tillage and habitat destruction; together with the removal of livestock and natural manure from crop rotations.

Earthworm numbers have also declined alarmingly owing to crop farming. Why should we be alarmed? Well, despite their humble appearance, earthworms give life to the soil, and soil is the basis of all other terrestrial life, including our own. Earthworms tunnel endlessly through the soil consuming and processing organic material as they go. What comes out of the worm's rear end enriches the soil and is the perfect food for plants. And as earthworms burrow their way through the soil they create its structure, with holes that provide aeration and allow water to drain through. Charles Darwin, the father of evolutionary theory, considered the earthworm to be the most influential animal in Earth's history. But long-term studies have seen half to all the worms disappearing, along with other soil animals and microorganisms. One review concluded that a change to organic farming with organic livestock manure was crucial to save earthworm populations. These studies also found that more than half the soil organic matter had been lost, and soil had become drier and more acidic. These are the first signs of fertile soils being turned to desert by intensive crop farming.

Pesticide is a catch-all term that includes insecticides, molluscicides, herbicides and fungicides. Pesticide usage is better documented in the UK than anywhere else in the world. In the UK, almost all pesticides (99%) are sprayed onto arable crops, fruit and vegetables. Less than 1% are used on grassland, in the rearing of cattle and sheep, even though we have more grassland than cropland in the UK.

Despite crop industry claims to have reduced pesticide applications to UK crops over the last couple of decades, government data shows that the reverse is true. The tonnage of pesticides applied has remained stable, but manufacturers have increased the toxicity of the active ingredients. Furthermore, the average number of active ingredients (toxins) applied to each hectare of UK cropland increased from ten in 1990 to over seventeen by 2014. Multiple toxins are now combined in each spray application, creating pesticide cocktails. For example, an average British wheat crop is now sprayed with over twenty toxins, combined in six or seven spray applications during each growing season. On average, thirty-one different toxins are applied to British potato crops, in sixteen spray cocktails each season. These cocktails often amplify the toxic effects on wildlife in unforeseen ways, as the active ingredients interact. For example, a 2018 study showed how combinations of active ingredients amplified the harmful impacts on earthworms. In California's almond groves, the honeybee mortality was exacerbated by interactions between fungicides and insecticides.

It is not just the effects of insecticides that impact on insect populations, birds and other animals. A 2019 Friends of the Earth report, *There's something wrong in the countryside: rising pesticide use in the UK*, describes how increased herbicide (weedkiller) applications, and particularly the increased use of broad-spectrum herbicides, is reducing the abundance and species diversity of non-crop plants. This in turn deprives insects, birds and mammals of habitat and food resources.

Together with pesticide use, habitat loss, owing to change of land use, is the most significant cause of species loss, even more so than climate change. The main drivers of this habitat loss are the conversion of natural vegetation and grassland to arable cropping and urban development. We have already seen in chapter 18 that soya, grown for human consumption, is a major driver of Amazonian deforestation. Cash-cropping (the growing of crops for sale or export, rather than for a farmer's own use) is also driving deforestation in Africa. In Britain, Europe and the US, conversion of grasslands to arable cropping is a major cause of habitat loss and species decline. A British study recorded significant population declines in 260 moth species between 1970 and 2010, for example, especially in areas where grassland had been converted to arable cropping or urban development.

In chapter 19, we saw how the conversion of grasslands in North America led to the 1930s Dustbowl disaster. But few lessons were learned, and from the 1960s to the 1980s there was further large-scale conversion of North American grasslands to crop production. Around 80% of the central grasslands have now been converted to cropland, and this has caused the decline of 74% of bird species. Loss and fragmentation of habitat, interference from cropping equipment and pesticides use were found to be the main causes. Historically, these diverse grasslands evolved with tens of millions of bison, elk and other ruminants, as well as the First Nations peoples who depended on their meat.

Those who champion meatless diets recklessly ignore what is staring us in the face: crop farming causes by far the most animal death and suffering and is also responsible for the ecological apocalypse we are witnessing.

Crop farming is inherently destructive to soil and wildlife. Without livestock to restore the soil and feed crops naturally, crop farmers are reliant on pesticides and fertilizers to maintain yields in the short term. But now we are beginning to glimpse the long-term costs. The plant-based dream of feeding a vegan world with no animals is simply a fantasy, and it is fast becoming a nightmare. Will we wake up in time?

We are told we can do many small things to save the insects, birds and voles: 'no-mow May' for our lawns, and never mind the blackfly on our roses or the slug holes in our hostas, and that we should leave the roadside verges to grow long and wild. But what about the endless fields of wheat, rapeseed and beet that cover nearly half our land? Never mind the small things, why don't we do one big thing?

Wheat is murder! Cows are the cure.

CHAPTER 21

EMBARGO THE AVOCADO

Food imports, food waste, water footprints and circular economies

'When we waste food, we contribute to the loss of biodiversity and the destruction of our ecosystems.'

Selina Juul

WHEREVER WE LIVE IN THE WORLD, A sustainable food system means eating food that is grown or reared as locally as possible.

Western consumers have come to expect a wide variety of food products on shop shelves all year round, especially exotic fruits, vegetables, nuts and other plant foods. Two thirds of fresh fruit and vegetables consumed in the UK are grown overseas, many of them thousands of food miles away. This import demand is partly driven by our distorted dietary guidelines, and by the growing number of non-meat eaters. The high dependence on imported food is one of the reasons why plant-based diets are so unsustainable.

A typical magazine article lists '15 Essential Ingredients for Vegetarian and Vegan Pantries' as beans and pulses (particularly lentils, chickpeas and their products, such as hummus); grains (especially rice, quinoa, millet and bulgar wheat); soya products (such as tofu, tempeh and miso); nuts (especially almonds and cashews); dried fruits (such as apricots and dates); tahini; dried sea vegetables (seaweed); coconut oil; maple syrup; and flaxseeds. Apart from flaxseed and some seaweed products, almost everything on this list is imported to Britain. Not to mention all the imported fruits and vegetables like sweet potatoes, bananas, citrus fruits and avocados. Let's take a look at a few other plant-based staples.

Legumes (beans, pulses, lentils, etc.) are a foundation of plant-based diets. The UK is now one of the largest importers of legumes in the world, with total imports valued at $155 million in 2020. Nearly 60% of these legumes come from developing countries where water is scarce – countries such as Kenya, Senegal, Zambia, Tanzania, Rwanda, Gambia, South Africa, Ethiopia, Egypt, Morocco, Jordan, United Arab Emirates, Guatemala, Peru, India, Pakistan, Burma, Malaysia and China. In addition, the UK imports $255 million of dried legumes, with Malawi being the fastest-growing import market.

Nuts are another vegan and vegetarian staple, and almonds are the most popular nut in the world. Americans eat more almonds than citizens of any other country, while Britain and the European Union are the biggest importers of almonds. The vast majority of these almonds come from California. As we saw in chapter 20, Californian almond growing is responsible for the killing of around 50 billion honeybees each year and countless other insects and wildlife. US almond growers use pesticides containing more than seventy active ingredients that are banned, or in the process of being banned, in the EU.

On top of the huge quantities of grains and seeds, quinoa, millet, spelt and bulgar wheat imports, the UK is also a major importer of rice ($644 million worth per annum), mostly from developing countries such as India, Pakistan and Thailand. And Britain is the world's top importer of chia seeds, which come mostly from India, Paraguay and Peru.

By contrast, the UK is largely self-sufficient in meat, milk and eggs. Self-sufficiency in lamb is 110%; some of the surplus is exported to France. Historically, there were significant imports from New Zealand, filling a seasonal gap in UK supply (sheep are spring breeders, so southern hemisphere lamb production mirrors that in the northern hemisphere). But New Zealand imports have plummeted as that country has found alternative markets in Asia, and UK production has filled the gap. The UK is 85% self-sufficient in beef (with most of the shortfall travelling only a short distance from the Republic of Ireland); 90% self-sufficient in poultry meat; and 65% self-sufficient in pig meat, with the shortfall travelling a relatively short distance from Denmark, the Netherlands and Germany; 100% self-sufficient in liquid milk; and 90% self-sufficient in eggs. Britain does of course import and export some dairy and meat products because of consumer demand for specialist products like French cheeses, but in general the food miles of livestock products are considerably lower than those of plant foods because our country is suited to growing grass and producing livestock.

Buying imported foods causes harm in many different ways, even if you do buy them from a local independent 'whole food' shop. In fact, whole food and health food shops are not that wholesome or healthy for the environment (or us). If you scan the shelves, you'll realize that they sell a very high proportion of imported products. Much of our imported food comes from water-scarce, climate-vulnerable and food-insecure countries. Much of it has a large water footprint and incurs high wastage rates during long transit times, owing to hot climates, poorer transport systems and less refrigeration.

Imported foods also have high food miles; the higher the food miles, the greater the greenhouse gas emissions from the fossil fuels used in transportation. Food transport emissions account for nearly 20% of global food sector emissions, and the highest proportion of these (36%) comes from fruit and vegetable transport. When food miles are expressed per kilo of protein, energy or other nutrients transported, fruit and vegetable nutrient-miles look even worse, and much higher than for meat, because around 85% by weight of this fruit and veg is water. The nutrient density of the remaining dry matter is much lower than meat.

It would be easier for the public to compare emissions caused by eating plant foods or meat if UN climate bodies such as the IPCC and the FAO presented the data in an equitable way. But as we saw in chapter 18, they don't.

YOU'VE GOT TO BE NUTS IF YOU BELIEVE BEEF COWS USE TOO MUCH WATER!

Water footprint is a term used to describe the amount of water that is used during the production and supply of a product, including food products. This should take account of all the water used from *field to fork*; not just the water used to grow the food in the field, but also the water used during cleaning and processing. The water footprint of a product depends not just on the amount of water used, but also on the source of that water. The footprint has three components, with differing impacts on the environment. Hence, the overall environmental impact of a product depends on the proportions of these three types of water used in its production.

The green water footprint is the amount of rainwater used. It is the water used in un-irrigated farming. It has the lowest environmental impact because the crops or grass are using rain that would have fallen on the land anyway. Most of the water is returned to the local water cycle by evaporation or transpiration from the plants, and through the urine and dung of the animals.

The blue water footprint is the amount of water we extract from rivers, lakes, reservoirs and groundwater. It is used for irrigating crops and to supply homes and businesses. Overuse, especially during dry weather, reduces groundwater reserves and the levels of rivers and lakes. This can lower the water table, affecting soils and leading to drought and harms to wildlife. Blue water use has greater environmental impact than the same volume of green water.

The grey water footprint is the amount of fresh water needed to dilute and assimilate pollutants, in order to meet water quality standards. Crops and animal production can have high grey water footprints if agrochemicals or manures run off the land, or leach through soils to pollute watercourses. Water extracted from polluted water sources has to be treated before use, often by dilution with blue water, so a large grey water footprint also means high environmental impact. (Water pollution also causes other environmental impacts such as emissions from the energy used during treatment, and toxic impacts on wildlife.)

Beef and other livestock products are frequently accused of having the highest water use, so let's look at these foods first. There is some truth in the claim where intensively reared livestock are concerned. Animals that are housed throughout their rearing require piped water. Most of that piped water is blue water extracted from surface or ground supplies, although green water from rainwater capture systems is sometimes used. Charles, the Lincolnshire farmer we met in chapter 16, supplied his entire herd of a thousand beef cattle with all their water from rooftop rainwater capture. Animals that are fed entirely on grains require large amounts of drinking water because of the dryness of grains. We also need to take account of the water used to grow the grains and other supplementary feeds, especially if the crop was grown with irrigation (blue water). Poor management of slurry and manure from housed livestock can cause water pollution and increase the grey water footprint (although it can be avoided with good management). This kind of intensive, high-impact water footprint applies particularly to pigs, poultry and intensive dairying. It also applies to intensive beef cattle that are housed and grain-fed, but only about 10% of the world's beef is reared in this way, mostly in North America, and they don't need to be.

When it comes to cattle, the water footprint very much depends on the intensity of the rearing system; as always, say after me: *it's the how not the cow!* This is because cattle are ruminants and have an advantage over monogastric livestock like pigs, in that they can live entirely by grazing grass. Grass has a high water content – about 85% depending on weather conditions and the stage of grass maturity.

In Britain and most of the world, beef and lamb are mainly pasture-reared, on grassland that relies on rainfall (green water footprint) and not irrigation. The cattle obtain most of their water requirements from the grass. They return most of their water intake to the pasture and the local water cycle via their urine and dung, and their dung increases the moisture-holding capacity of the soil. Even in hot dry weather, any extra (blue) water that cattle take from field troughs or rivers is also urinated back to the land. Pasture-reared beef and lamb also typically have a very small grey water footprint, because agrochemical use is so low. Only during winter housing do cattle use much piped (blue) water, but this is generally at times when rainfall and reservoir levels are high, in contrast to crops, which need most water at the driest times of year. As a result, pasture-fed cows and sheep have the smallest, greenest water footprints *of any foods.*

By contrast to beef cattle and sheep, grains, vegetables, fruits and nuts require a lot of water to grow, and have large water footprints. In many parts of the world from which we import them, those crops have to be irrigated using lots of high-impact blue water. Even in eastern England – Britain's most important crop-growing region and lowest rainfall area – many crops have to be irrigated. Crop farming generally can have a high grey water requirement as a result of agrochemicals and soil erosion that cause water pollution.

Whether fruit and vegetables are grown with green water or blue water, importing them means translocating large quantities of water around the world, often from areas of water scarcity, climate vulnerability and greater food insecurity, to countries like Britain which have greater water security, lower climate vulnerability and greater food security. This transfers large volumes of water from the exporting region's local water cycle to our local water cycle. And changes in local water cycles play an important part in climate regulation. As we have already discussed, the water cycle is important in temperature regulation as well as precipitation, so perhaps we should be wondering how big a part the global food trade is playing in climate change.

The proportion of fruit and vegetables supplied to the UK market from climate-vulnerable countries increased from 20% in 1987 to 32% in 2013. During the same period, the proportion of UK-grown fruit and vegetables supplied to the domestic market fell from 42% to 22%. Legumes (beans and pulses) showed the largest increase in supply from climate-vulnerable production areas, rising from 0.2% to 19% of supply between 1987 and 2013. Approximately 76% of the fresh water consumed in the supply of fruit and vegetables to the UK is drawn elsewhere, including from countries with a high risk of water scarcity such as Spain, Egypt, South Africa, Chile, Morocco, Israel and Peru.

And if you think avocados are a superfood, think again. UK and global avocado demand has spiralled in recent years. Most UK imports come from Peru, South Africa, Chile, Israel and Spain – water-scarce countries where erratic rainfall and drought are major concerns. Around 25 billion litres of water are used in these five countries each year just to supply the UK demand. In the Petorca region of Chile, extreme water shortages have meant water restrictions of 50 litres per person per day (about a third of what we typically use in Britain), while avocado growers continue to irrigate each tree with more than this quantity of water. The escalating Western demand for avocados is also causing many other impacts, such as in Mexico (the world's largest producer), where the crop is causing rapid deforestation, water pollution and health impacts.

And finally, the prize for the biggest global water footprint of any food group goes to those staples of the plant-based diet – nuts. Almonds and cashews are the worst of all, using more blue water than any other crops.

FOOD WASTE

Globally, we waste 1.6 billion tonnes of food every year. The highest level of waste occurs in the fruit and vegetables sector (40–50%), followed by fish (35%) and grains (30%), oilseeds (20%), then meat and dairy (20%). In Europe and North America, we waste over 100kg of food per person per year. The UK wastes almost 10 million tonnes of food every year, especially potatoes (5.8 million tonnes) and bread (900,000 tonnes), and 40% of food waste in the UK hospitality industry is carbohydrates. Most domestic food waste consists of vegetable matter and mouldy bread, rather than meat or fish. In industrialized countries the greatest proportion of the food waste occurs at retail level due to produce that does not meet standards of appearance or deteriorates or does not sell before the sell-by date. In developing countries most of the waste is from deterioration during transport, which is another reason why buying imported foods is so unsustainable.

This food waste matters because it has serious environmental and human costs. Some 28% of the world's agricultural land (1.4 billion hectares) is wasted each year, along with enough water to fill three Lake Superiors, growing food that is never eaten. The waste food also produces 7% of global greenhouse gases, equal to the total emissions of India, the world's third-highest emitting country.

The global population is expected to increase from 8 billion currently, to 9.7 billion by 2050, and there are already 768 million undernourished people, mainly in Asia, Africa and South America. Think tanks such as the World Resources Institute (WRI) suggest we will need to find 600 million extra hectares of agricultural land to meet the global food gap. The risk is that this land will be found at the expense of further deforestation or conversion of grassland to crop farming, increasing biodiversity loss and emissions. However, if we reduced global food wastage by less than half, we wouldn't need this extra arable land. As we have seen above, reducing food wastage primarily means reducing plant-food wastage and the global trade in plant foods.

Instead, however, the think tanks focus on reducing meat consumption, which they claim will release land for more crop production. This might be the case if we reduced pork, chicken and egg consumption, because these livestock are dependent on grains. But instead the WRI suggests that beef and lamb consumption should be halved while 'improving the marketing of plant-based foods, improving meat substitutes and implementing policies that favour consumption of plant-based foods'. As we've mentioned before, most beef and lamb is not reared from land that can grow crops.

Centred in the crosshairs of these think tanks and global fixers are the world's grasslands. These cover more than half the world's utilized land, from the central grasslands of North America, to Britain's hills and uplands, to the central Asian steppes and the African savannah. Globally, these grasslands are home to 1.3 billion pastoralists and millions of small farmers, who use low-input, low-impact grazing methods to pasture-rear livestock. These grasslands and their pastoral peoples are deeply misunderstood and undervalued by wider society. Pastoralists are frequently seen as inefficient land users who should be forced into more intensive sedentary agriculture or urban livelihoods, to free the lands for crop farming, mining or other uses. This fails to understand that these lands are grassland because they are too marginal and sensitive to be cropped. When allowed to follow their traditional herding patterns, the pastoralists show great expertise in reading the environment to optimize the land's productivity without damaging it.

Our widely held condescending attitudes towards grasslands and pastoralists also fail to recognize that they are highly important ecosystems in themselves, providing biodiversity, carbon storage, and healthy water cycles. Converting more of these grasslands to cropping or other uses will compound our environmental crisis. We have already seen the impacts during the Dustbowl disaster and later conversion of North American grasslands.

Livestock are vital for circular economies. Crops produce a lot of waste and by-products, such as straw, rice husks, bran, waste vegetables, brewers' grains and oilseed residues. Many of these are fed to livestock. Doing so means livestock reduce food waste and increase productivity in our food systems, as explained by the UN FAO report 'Livestock Solutions for Climate Change'. Crop wastes and by-products make up around 30% of livestock feeds worldwide, and the FAO notes that 'in Japan, 52% of waste from the food industry is now used as livestock feed'. Ruminants like cattle are especially important, as their specialized digestive tract enables them to get the most out of even low-quality crop by-products, like straw, and upcycle them into high-quality meat and milk. In the intensive crop-growing areas of eastern England, thousands of sheep are brought in each autumn to clear up thousands of hectares of crop waste that would otherwise rot in the fields or be ploughed in using lots of diesel fuel, producing CO_2, methane, smell and flies. Cattle are also fattened on waste carrots, potatoes and other vegetables that do not make the grade for human food.

Clearly, when it comes to choosing a sustainable diet, the assumption that 'meat is bad, plants are good' is far too simplistic. We have to consider all the impacts throughout the whole supply chain, from farm to fork. Consuming locally produced food – especially pasture-reared meat – will do far more to reduce these harms than changing to a plant-based diet. Britain's land and climate is ideal for cattle and sheep rearing at pasture. We already produce most of our domestic beef and lamb requirements. If we put grass and grazing livestock back into our crop rotations, we would comfortably meet our needs, replace intensive livestock and reduce crop agrochemical use.

Maybe you are asking how we expect people to afford to eat the Sapiens diet. After all, meat is expensive, isn't it? And as we've already explained, we are not suggesting everyone should eat intensively reared chicken and pork – quite the opposite. But Sapiens isn't about eating huge plates of fillet steak three times a day. Beef fat, liver and other organ meats are cheap, and there are many cheaper cuts of meat, like braising steak and mince, that are just as healthy and nourishing.

You'll be surprised to find that eating the Sapiens way without all those carbs to drive insulin and hunger is very satisfying; so much so that you'll almost certainly feel like eating only twice a day, with no expensive snacks in between. You'll be even more surprised if you tot up the cost of all those fruit and veggies, bread, pasta, rice, beans, jams and breakfast cereals that you currently eat – especially considering how much of it goes in the bin each week! And if you work out the cost of a week of healthy, nutritionally balanced Sapiens meals compared with vegetarian or vegan meals meeting the same nutritional criteria (with added vitamin supplements of course), you'll be surprised to notice that your bank balance is healthier, as well as your heart.

The Sustainable Food Trust's 2022 report, *Feeding Britain from the Ground Up*, states,

> Millions of people are now asking, 'what should I eat to be healthy and sustainable?' ... our response is that we should align our diets to what the UK can sustainably produce, because without changing what we eat, it will be impossible for farmers to switch to sustainable production methods and remain economically viable.

The report sensibly proposes that we in the UK should at least maintain current levels of beef and lamb production, as well as putting grass, cattle and sheep back into crop rotations to reduce crop farming's reliance on agrochemicals and increase organic production. The SFT recommends reducing pig and poultry production and thereby reducing the land needed to grow grains for those intensive livestock. The land freed up would enable us to increase our self-sufficiency in fruit, vegetables and nuts (though consumers would need to adjust their taste for exotic produce to temperate varieties). But why not go a step further? Why not cut the amount of grains we grow even further, along with oilseeds and sugar – the grains, oilseeds and sugars that go into human foods and cause so much of our ill-health? This would free up even more land for pasture-reared meat, as well as more fruit and vegetables – if you really still want them after reading this book!

And for readers in sunnier climes, the principles remain the same. Eat what you can produce locally, sustainably. Wherever you are in the world, extensively reared grazing livestock like cattle, sheep and goats have a big part to play in your sustainable food system, providing restorative manure for crop rotations, reducing soil erosion and desertification, recycling crop wastes to provide a circular economy and providing nutrient-dense, healthy food without costing the Earth.

* * *

Food is a necessity. Farming, using soil and sunlight to produce food, is also a necessity. In Section 2 of this book we saw how diet impacts our health, and now we've seen how farming, to provide our diet, affects our planet.

Diet and health, farming and the environment; four elements that are inseparable.

Recognizing this and rejecting modernist ideas about synthesizing food in laboratories and factories (high technology has landed us in enough trouble already), we reach our conclusion: that we should return to former, older, wiser patterns of eating and farming to improve our health and safeguard our planet.

Conclusion

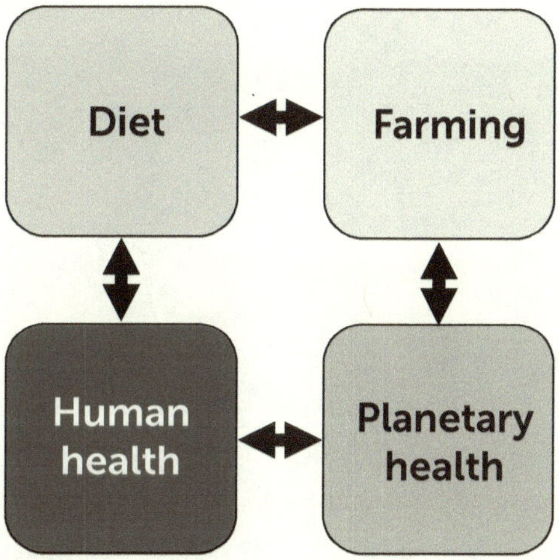

CONCLUSION

OLD SOLUTIONS FOR MODERN PROBLEMS

'Making food cheaper has not made people better off, financially or otherwise. The solutions, it seems, may not be new ones. They can be found in books published a hundred years ago.'

Sarah Langford, *Rooted*, epilogue

LET'S SQUEEZE THE LAST 300,000 YEARS INTO a single day. In this timeframe, every hour represents 12,500 years. For over twenty-three hours *Homo sapiens* thrived on animal foods. It was only at 11:12 p.m. that we adopted farming. And since starting to farm, which increased our consumption of plant foods, our brains have started shrinking. A mere thirty seconds to midnight heralded the inclusion of vegetable oils in our diet. Around the same time agrochemical crop farming escalated soil depletion and ecosystem collapse. And it's only been in the last ten seconds that the American and UK dietary guidelines have been in place – but ten seconds of soaring obesity, diabetes and neurological disorders.

If we carry on this way, who knows what the next ten seconds will bring?

* * *

Ten seconds. On our clock that's thirty-five years. Thirty-five years which would be well spent tackling obesity, diabetes, biodiversity loss and climate change. What we eat and what food is produced are interlinked, but nevertheless this link is nearly always ignored by dieticians and environmentalists. Dieticians blithely suggest changes to diet without a thought as to what's involved in supplying that food. And climate change experts make recommendations for farming with little appreciation (merely lazy assumptions) of the impacts that these may have on food supply and consequently our health.

But it so happens that more sustainable farming and healthier eating can go hand-in-hand. Maybe this isn't just good fortune. By returning to an ancestral diet as we propose, it could be that we also return to a natural, evolutionarily proven way of feeding humans that suits both our well-being and the health of the planet. This may sound naively utopian; as Chris Smaje put it in his book, *A Small Farm Future*, '[the] great difficulty is that modernist notions of progress cut the ground from beneath the feet of anyone who tries to articulate past solutions to present problems'. But we believe it's worth aiming for.

WHAT WE HAVE LEARNED IN THIS BOOK

An animal-based diet can be healthy, environmentally friendly and less cruel.

Our evolution as a species was driven by first scavenging and later hunting a diet that comprised meat, animal fat and offal. These foods provided the vitamins, minerals and fats that led our brains to triple in size over the last 6 million years. Over this period of time the human digestive system has adapted to expect this 'ancestral', animal-based diet. Our ancestral diet remains the optimum diet for us today. We are the only species on the planet that chooses not to eat the foods it evolved to eat.

Ignoring evolution, and misinformed by flawed dietary studies, today's nutrition guidelines recommend a diet that is inadequate in nutrients yet contains excessive amounts of carbohydrates in the form of grains, legumes, fruit and vegetables.

As to the cost of the Sapiens diet, offal and animal fat – the essential and most nutritious parts of the diet – are ridiculously cheap. Per nutrient, fruit, vegetables, bread and pasta are the most expensive foods. And these expensive foods are also the ones that end up in the bin – every year we waste 28% of the world's agricultural land growing food that is never eaten – and most of this food waste is plant foods. Just halving food wastage would enable us to feed the world's growing population, without further destruction of forests, grasslands or wetlands.

Having established that animal foods are what we should be eating, we can also anchor the health benefits of the Sapiens diet with its environmental and ethical benefits. After all, what would be the point of having a healthy diet that damages the planet and causes more animal suffering?

Current climate change recommendations for farming ignore or misunderstand the implications for diet and the environment. They typically assume that livestock farming can be eliminated with no negative consequences. But if we eat less meat, what do we eat instead? Yet more carb-heavy, nutrient-poor plant foods, leading to more obesity, more diabetes and more mental health disorders, which in turn lead to more drugs that end up in our water, harming ecosystems. And if we get rid of grazing livestock, how will we grow crops? With even more agrochemicals, soil loss and greenhouse gas emissions.

Our farmers – the people who played such a major part in the survival of the nation during the 1940s and 1950s – have since been frustrated by decades of confused and contradictory farming and environmental policy. They have unfairly become the whipping boys of environmental activists, as they have struggled to negotiate the muddle of policy, bureaucracy and market forces.

We must also wake up to the cultural, societal and environmental devastation that these attitudes and policies towards livestock and meat are causing globally, such as the marginalization and dispossession of millions of traditional pastoralists, as well as small farmers and entire rural communities. Even as we make final edits to this book great injustices are occurring in the name of *the environment*. In Tanzania, 82,000 Maasai people are being forcibly transferred hundreds of miles from their ancestral lands in the Ngorongoro, as part of government policy on nature conservation and (environmental) tourism.

Grazing livestock do not waste land; they add to the global food table by upcycling vegetation that we can't eat, from land where we can't grow crops, into the most nutritious human foods. Grasslands would not serve the world better if rewilded or if ploughed up to grow yet more crops; and repurposing grassland to plant new trees should be regarded as environmental vandalism.

Crop farming kills billions of birds and mammals every season, along with quadrillions of insects, worms and other tiny life forms that all life ultimately depends upon. Eating plant foods actually causes far more animal death and suffering than eating pasture-reared meat. *All* crop growing causes some degree of environmental harm, while only *some* kinds of livestock farming are harmful – intensive, grain-fed livestock farming. Even conventional British beef and sheep farming is low-input and much less damaging than arable crop production. It would require very little adaptation to make this positively beneficial, if there were the political will to support wide transition to regenerative or organic, pasture-reared livestock farming. And only by including livestock in crop rotations can we mitigate the harms of arable farming. When it comes to livestock farming, what matters is *the how not the cow.* When it comes to plant-based farming we should *ditch the plough and add the cow.*

We need to rethink our farming and food systems; we need a more local approach. Wherever we live, we should eat local and reduce imported foods. Grazing livestock are reared all over the world and make an irreplaceable contribution to human nutrition globally. Britain, like many other temperate countries, is well suited to producing meat and animal foods. Despite this we are now heavily reliant on environmentally harmful imported plant foods.

The current system promotes unhealthy food and environmentally destructive crop farming. We need a new paradigm that properly values organic and regenerative grassland farming practices that sequester carbon, encourage biodiversity and also provide nutritious, healthy food.

WHAT CAN WE DO ABOUT IT

Diet must come first – not least because the environmental costs of poor health are much higher than most realize. We can then organize our land use and farming systems to deliver that diet, not just trying to mitigate environmental harms, but actively building health into the environment.

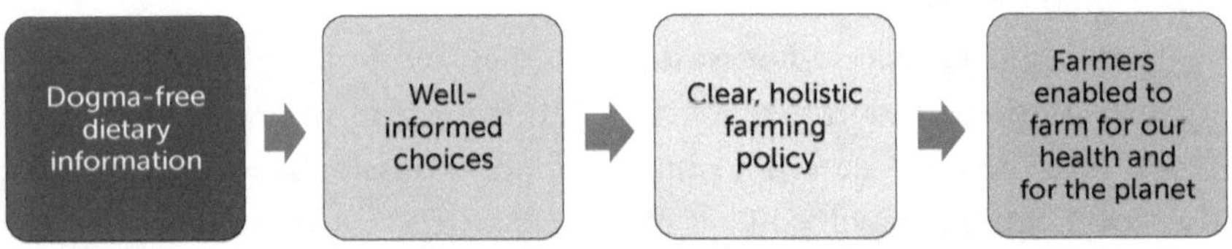

The tricky part, of course, is convincing those in positions of power, authority or influence to put these objectives into effect. We hope that this book will make a difference. But many have gone before us – Jayne Buxton, Mark Nathan Cohen, Loren Cordain, Georgia Ede, Zoë Harcombe, Sarah Langford, Staffan Lindeberg, Nicolette Hahn Niman, Diana Rodgers and Robb Wolf, Paul Saladino, Gary Taubes, Nina Teicholz and Marina Young, to name more than a few. Yet still the pyramid remains stubbornly upside down, still we are urged to base every meal around carb-heavy plant foods, still the media report that animal fats cause heart disease and that plant foods are nutritious, and still the myths persist that livestock farming does nothing but pollute while crops somehow, magically, are grown with no negative impact on the environment.

OUR KEY MESSAGES

DIET AND FARMING ARE LINKED

Since they are linked, you can't change the policy for one without altering the other. This needs to be recognized by dieticians, environmental researchers and policy makers.

DIET AND HUMAN HEALTH

We can improve our health by making the majority of our diet animal-sourced foods – that is, meat, fat and offal. Consequently, there must be a significant reduction in plant-sourced foods, especially grains and vegetable oils.

FARMING AND PLANETARY HEALTH

We can supply the Sapiens diet through environmentally friendly, regenerative livestock farming.

WE END WITH A FRESH START

It's clear to us that many of the threats to our health and to planetary health stem from crop farming, almost all of which is intensive and agrochemical-based, destroying the very soil that all life depends on, and producing food that is a major cause of modern, 'Western' diseases and obesity. The solution is equally clear: we should eat to meet our nutritional needs, from locally produced, regeneratively farmed animal foods.

Having made our case, we recognize that not everyone will read this book and immediately adopt the Sapiens diet. Everyone needs to make up their own minds about what they want to eat. Young children, though, can't make up their own minds – they eat what is given to them. One of our main motivations for writing this book is the potential irreversible harms that occur to infants and children that are fed a plant-food only diet. With children, especially, we need to avoid such diets and nourish our future generations to give them the best chances in life. Babies should be weaned onto animal foods, not baby rice and carrot purée.

Another driver for this book was our vision of an adult population that remains physically and mentally healthy throughout later life, and a society free of the immense burden of metabolic disease.

But to move away from intensive plant-based foods towards pasture-fed animal foods and crops grown organically with livestock in the rotations will, if we're honest, require more than convincing the readers of this book. What could make a substantial difference would be a change in the attitude of governments and international organizations, such as the IPCC, FAO and WHO. And, of course, the media. Too often the national newspapers and big-circulation magazines are full of baloney about diet, and badly researched articles that are more about colourful pictures and celebrity lifestyle choices than they are about sensible dietary advice. They publish food stories about 'how to lose weight', about vitamin D and iron deficiency, and avoiding mental decline, but never once mention the value of meat, animal fat and offal, instead recommending supplements and harmful low-fat diets.

This book gives us all the accurate, agenda-free information we need to make informed choices about what to eat. We need to rewrite dietary guidelines to encourage animal-sourced foods and discourage plant-food consumption, and create a demand for healthy food that farmers can produce without costing the Earth.

From this perspective, we reach our conclusion. Deviating from our natural diet has caused untold harm to our health and to the health of the planet. The Sapiens diet is the diet evolution designed for our species. On the ground this means changing current dietary guidelines and go so far as to introduce warnings of the health dangers posed by grains, carb-heavy foods, legumes and vegetable oils on food packaging, as per tobacco. For food production, we need policies, such as taxes on agrochemicals, to discourage intensive arable and animal farming and to incentivize organic, regenerative and pasture-fed livestock farming. We also need to properly reward and incentivize organic and regenerative farmers, so that they can afford to produce food that nourishes us and the environment. Globally, we must protect grasslands and pastoral livestock herders. And we must disincentivize the consumption of imported foods that have been produced and distributed in ways that damage the environment.

We cannot provide for 8 billion people and rising on crops that are destroying soil, water supplies and ecosystems. If we persist with the flawed logic that we should replace meat with plant-based diets, then environmental destruction will continue to escalate, along with the socio-economic and environmental costs of diet-related disease.

By returning to basics, eating and producing the animal foods that our bodies evolved to eat, all the benefits follow. Within days of simply changing what we consume, we can rectify insulin levels and leaky gut – the commonest causes behind most Western diseases.

Unlike governments, we are not telling you what to eat or how to farm. We are informing you about the best ways to eat and farm; it is up to you what to do with this information. But we must anchor the health benefits of the Sapiens diet with sustainability. We are being realistic, not dogmatic: offering sound information on which you can make your own choices.

* * *

The dawn of agriculture heralded what we call civilization. Arable farming supported dense populations and the growth of cities. It provided calories for the people. But for human health it was a disaster. We sleepwalked into a class Z diet, brimming with carbohydrates but barren in nutrients. Arable farming deprived us of our optimum diet and optimum health, and, having done that, is now helping destroy the planet.

But let's not wait for new government policies on sustainable farming, or for health authorities to give up groupthink and revise dietary guidelines – it could be a long time before those pigs fly. Rather, we can as individuals take the initiative and decide by ourselves, to whatever degree we can, to pursue a healthy and sustainable diet.

Eat as our distant, or indeed not so distant, ancestors ate,
on locally produced, animal-sourced food, nurtured regeneratively.

Live longer, live better and tread lightly on the Earth.

APPENDIX A

TRAFFIC LIGHT SYSTEM FOR FOODS, WITH THEIR EFFECTS ON BOTH HEALTH AND THE ENVIRONMENT

Here's a summary that draws together the health and farming pros and cons associated with the dietary recommendations we are making in this book.

Diet		**Why**
• Animal fats • Pasture-reared beef and lamb • Offal, e.g. liver, liver pâté	✓✓	• Natural, ancestral diet, to which we are evolutionarily adapted • Full range of nutrients in correct amounts for body and brain • Pasture-reared is environmentally friendly and climate-positive
• Fish, poultry, eggs, dairy products, fruit, vegetables, nuts • Processed meats containing additives, e.g. nitrites and sulphites • Artificial sweeteners	✓	• Fish, poultry and eggs good for health but less good for environment • Eat fish sparingly, following the Good Fish Guide or choosing Good Catch restaurants • Fruit, vegetables, nuts and dairy are not ideal for health but better than grains and vegetable oils • Fruit and vegetables nutritionally incomplete and contain anti-nutrients • Most crops are intensively farmed and damage the ecosystem and environment • Nitrites, sulphites and artificial sweeteners may be harmful to health
• Sugar • All grains (wheat, rice, oats, barley, rye) • Legumes (such as peanuts, pulses, peas) • Vegetable oils	✗	• High intakes of sugar and carbohydrates associated with diabetes, obesity, cardiovascular disease; implicated in cancer and dementia • Grains and legumes – full of carbohydrates and anti-nutrients; harmful to health; cause leaky gut, diabetes and obesity; implicated in autoimmune diseases • Vegetable oils – incorrect balance of fats, cause leaky gut and inflammatory diseases • Intensively farmed crops damage the ecosystem and environment

APPENDIX B

MEAL IDEAS AND OTHER PRACTICAL TIPS

WHEN A FEW YEARS AGO ONE OF the authors learned that a friend had been diagnosed with type 2 diabetes, they were dismayed when the diet that had been recommended to help control the condition was described. Reducing sugar was to be expected, such as in cakes, muffins, desserts and other sugary snacks. But refined carbs were out as well – breakfast cereals, croissants, bread. *And bread?* Surely not! No toast with breakfast, no lunchtime sandwich, no bread and cheese in the evening while watching *The Great British Bake Off*. No hamburgers, no hot dogs, no bacon rolls. It sounded like a joyless existence.

Yet now, the authors opt to not eat all those things, and out of choice, not necessity. And it's far from joyless.

We realize that it is unrealistic for most to stick to the Sapiens diet 100% of the time – even we don't manage it. With this in mind, here are some meal plans which include low-carb, paleo, keto and Sapiens diet suggestions. Obviously, aim for Sapiens meals as much as you can but low-carb, paleo and keto meals are not bad runners-up.

Before embarking on any of these diets, if you are on any medication or suffer from any medical condition, consult a specialist (preferably one with experience of keto diets) before making any changes. Remember, these diets are as powerful as drugs – just as you wouldn't start a new drug without consulting a doctor, so you shouldn't start these diets without consulting a specialist if you suffer from any medical condition. These diets lower blood glucose, blood insulin and blood pressure dramatically within days of starting the diet. If you are on medication for any of these conditions, then your blood glucose and blood pressure could drop to dangerously low levels if your medication isn't adjusted when starting the diet. And owing to the fact that many psychiatric conditions improve too, medication for these conditions might also need adjusting. The good news is that you might eventually be able to stop all your medication.

There are many resources to help you along the way. Some of the best ones are:

- Georgia Ede, *Change Your Diet, Change Your Mind*
- dietdoctor.com has many, many recipes
- diabetes.co.uk: www.diabetes.co.uk/reversing-diabetes.html

When embarking on one of these diets, a sensible path to take would be to start off low-carb, and then progress to paleo, then keto and then the Sapiens diet. Allow yourself at least one month on each diet. And if you want to track any markers to monitor changes, here are a few suggestions.

- If you are overweight, check your weight once a week (first thing in the morning before having anything to eat or drink).
- Blood pressure, blood glucose, leaky gut test (zonulin or PEG 400 test), Kraft insulin test, waist to height ratio (the lower the better), blood triglyceride level.
- It would be worth making a list of minor ailments and/or any skin conditions that you might be suffering from before embarking on any of these diets. And take a photo of yourself before – many skin conditions also improve quite quickly.

Here is a list of common conditions that often get better:

- Acne
- Dry and sensitive skin
- Eczema
- Heartburn
- Dermographism
- Dark circles around eyes
- Sinus congestion
- Sneezing
- Wheezing
- Coughing
- Bloating
- Constipation
- Diarrhoea
- Reflux
- Fatigue
- Pigmentation on face, elbows, knees and armpits (known as 'acanthosis nigricians')
- General aches and pains

At the outset, it is a good idea to clear out your kitchen of foods that you shouldn't be eating. Have a final binge, or give them away, or chuck them.

If you do manage to follow the Sapiens diet most of the time, you should feel full enough to have only two meals a day and not need to snack at all. We usually skip breakfast and have lunch between noon and one p.m. and then dinner between six and seven p.m., even on days when we don't keep to the Sapiens diet.

One of the most important things to remember is to include some offal in your diet at least twice a week – the best being brain, bone marrow and liver. These are the most nutritious foods of all. If you can't stomach these in unadulterated forms, then liver pâté is a good alternative. Or try adding small amounts of minced offal to meat burgers. Bone marrow fat does not have a strong taste and can be added to gravy or used as a cooking oil.

Prioritize offal in your diet. Instead of trying to include some fruit and vegetables every day, include a bit of liver or liver pâté instead. (We usually have a couple of pieces of fried liver with our brunch every morning, to get it over and done with at the beginning of the day.)

If you find the taste of liver too strong, try cooking it with garlic, salt, pepper, paprika and butter or cream (this is similar to the liver that Nando's sell, and is delicious). A good friend adds a small piece of dark chocolate to her liver recipe …

A FEW COOKING HACKS

Nearly all cuts of meat can be cooked until they are meltingly tender in a slow cooker. We think it is worth investing in a slow cooker – it can stay on even when you aren't around and is safer than leaving an oven on unattended.

If you are into making bone marrow broth, skim off the fat and use it as a cooking fat – it keeps in the fridge for a couple of days and in the freezer for a couple of months. After boiling the bones, wait until the broth has cooled in a fridge and you can just lift off the fat layer quite easily. Most of the nutrients from bone marrow are in the fat – it is incredibly nutrient-dense – so don't throw it away, eat it!

Ask your butcher for beef fat trimmings and make your own beef dripping (many butchers will give it away for free). Heat the fat gently for an hour or so, strain the melted fat and store in jars in the fridge.

Don't discard fat from any cuts of meat – cook them and eat them! A good way of making fat easier to eat is to cut it into small bits and fry. For example, the fat on the sides of steaks such as sirloin steaks can be done in this way.

For keto recipes use beef dripping or ghee or butter instead of cooking oil.

For Sapiens recipes use beef dripping or lard or bone marrow (skimmed off bone broth) instead of cooking oil.

If slow cooking anything, make enough for two days or even more and freeze some for another day.

If you can't be bothered with cooking, most of the big supermarkets (Iceland, Waitrose, M&S, etc. …) now do excellent, slow-cooked ready meals. They don't usually have any unnatural additives in them – just meat with herbs or spices and a sauce.

Cow sharing. Get together with a group of friends and you can buy a whole cow together – this works out cheaper than buying beef in a supermarket.

GENERAL TIPS

Buy meat and eggs from a local farm (if you can't make it in person, many farms sell direct to customers online).

- Browse websites such as the Ethical Butcher – they sell regeneratively farmed meat.
- Try and source pork and eggs from farms that breed outdoor pigs and chickens in good conditions.
- Avoid eating raw vegetables – they have higher levels of anti-nutrients than when cooked.
- Do bear this hierarchy in mind when planning your diet and the meals to go with it:

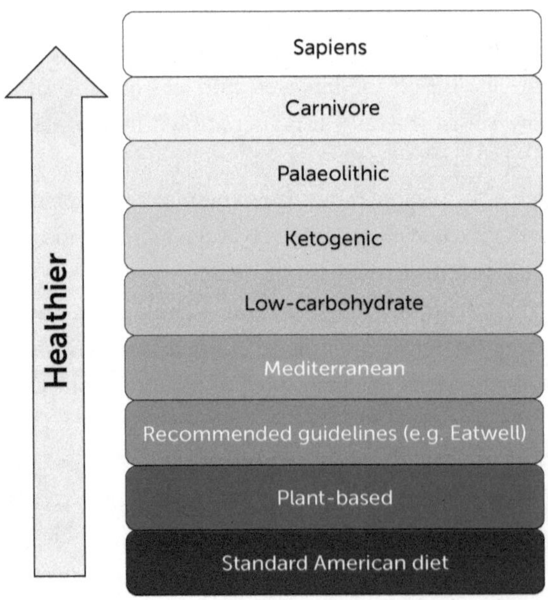

BREAKFAST/BRUNCH/LUNCH OPTIONS

- Liver with garlic, salt, pepper and paprika in beef dripping (*low-carb and keto, Sapiens if no garlic or paprika*)
- Fried, scrambled or poached eggs (*low-carb, keto, Sapiens*)
- Eggs royale, Benedict or Florentine without a muffin (*low-carb, keto*)
- Eggs royale, Benedict or Florentine with a gluten-free muffin (*paleo*)
- Plain omelette (*low-carb, keto, carnivore, Sapiens*)
- Omelette with a filling: cheese, ham, vegetables, salmon, etc. (*low-carb, keto*)
- Classic bacon and eggs (*carnivore, low-carb and keto if cooked in ghee, butter or vegetable oil, Sapiens if cooked in beef dripping*)
- Beef, lamb or chicken liver pâté (*low-carb, keto, carnivore, Sapiens*)

It doesn't take long to get used to having bacon and eggs or any kind of breakfast without toast or any other carbs. If you need something 'carby' while transitioning, have potato or parsnips in one form or another.

POTATO AND PARSNIP OPTIONS

- Roast parsnips or roast potatoes in dripping, hash browns, rosti, potato mash, parsnip mash, potato waffles, jacket potatoes, boiled potatoes, potatoes Dauphinoise, potato gratin, homemade chips (in dripping) or frozen chips.

All the potato options can be made from scratch or bought readymade.

Potatoes are a 'nightshade' vegetable that contain alkaloids which can be a problem for people with autoimmune conditions. It's best to avoid eating potato skins, as they have the highest concentration of alkaloids and other anti-nutrients.

LUNCH AND DINNER OPTIONS

- Meatballs: lamb, beef or venison (*low-carb, paleo, keto, carnivore, Sapiens*)
- Meatball curry (*low-carb, paleo, keto, carnivore, Sapiens*)
- Meatballs in marinara sauce (homemade or readymade) (*low-carb, paleo, keto, Sapiens*)
- Lamb kofta (*low-carb, paleo, keto, carnivore, Sapiens*)
- Lamb kebabs (*low-carb, paleo, keto, carnivore, Sapiens*)
- Slow-cooked shoulder of lamb with korma sauce (homemade or readymade) (*low-carb, paleo, keto, Sapiens*)

- Slow-cooked shoulder of lamb with gravy (*low-carb, paleo, keto, carnivore, Sapiens*)
- Slow-cooked shoulder of lamb with mint sauce (*low-carb, paleo, keto, Sapiens*)
- Coddled eggs with smoked haddock and spinach (*low-carb, keto*)
- 'Spaghetti' bolognese (use a spiralizer to make 'spaghetti' from courgettes, or some supermarkets sell it readymade) (*low-carb, paleo, keto, Sapiens*)
- Slow-cooked ribs: plain, with gravy, or glazed with barbeque, honey or date sauce (*low-carb, paleo, keto, Sapiens*)
- Beef goulash (*low-carb, keto*)
- Pork belly, slow-cooked with pork crackling, sweet date sauce or apple sauce or cider jus (*low-carb, paleo, keto, Sapiens*)
- Steak. Try cutting off any side-fat, cube it and then fry alongside the steak (*low-carb, paleo, keto, carnivore, Sapiens*)
- Rack of lamb (*low-carb, paleo, keto, carnivore, Sapiens*)
- Leg of lamb with mint sauce (*low-carb, paleo, keto, Sapiens*)
- Pork shoulder (slow-cooked) (*low-carb, paleo, keto, carnivore, Sapiens*)
- Shepherd's pie (*low-carb, paleo*)
- Roast chicken. Cover in a muslin cloth that has been soaked in melted butter, then roast in dripping or goose fat with mixed herbs, garlic, paprika and onions (*low-carb, paleo, keto, Sapiens*)
- Steamed mussels in white wine sauce (*low-carb, keto*)
- Stir-fried prawns with garlic and chilli (*low-carb, paleo, keto*)
- Pan-fried salmon or sea bass, with optional dollop of butter and/or poached egg on top (*low-carb, keto*)
- Pastry-less quiche (traditional quiche Lorraine, or quiche with salmon or with vegetables. Just throw away the pastry, or if cooking it yourself, make it in a pie dish without pastry. Some supermarkets sell 'crustless' quiche – simply discard the pastry base) (*low-carb, keto*)
- Pasta-less lasagne. Use pan-fried courgettes or carrots instead of pasta, and gluten-free flour for the white sauce. Try adding garlic and a touch of paprika when cooking the mince (*low-carb, paleo, keto*)
- Moussaka (*low-carb, paleo*)
- Cottage pie (*low-carb, paleo*)
- Twice-baked cheese soufflé (use gluten-free flour for the white sauce base) (*low-carb*)
- Smoked salmon or mackerel (*low-carb, paleo, keto, carnivore, Sapiens*)
- Salmon terrine or mackerel pâté (*low-carb, keto*)

VEGETABLE OPTIONS (IF YOU MUST)

- Mushrooms in butter, cream and parmesan with mixed herbs and a touch of paprika. (Cook the mushrooms over a low heat in butter or ghee until all the fluid from the mushrooms has dried out, fry the garlic in butter or ghee, add the fried garlic, herbs, cream and parmesan to the mushrooms)
- Spinach with garlic and butter, with or without cream, or with cream and cheese (such as Gruyère or parmesan)
- Cauliflower cheese (use gluten-free flour for the white sauce)
- Cauliflower rice
- Veggie 'noodles' or 'spaghetti' (spiralized vegetables – looks like spaghetti or noodles – courgette, butternut squash, carrots)
- Coleslaw (chopped cabbage and carrots with mayonnaise; but beware uncooked vegetables and also the vegetable oil in the mayonnaise)
- Low-carb vegetable salads

Depending on which vegetables and how much of them you eat, these are okay for all the diets except for the carnivore diet and for the strictest version of Sapiens.

SAUCES

- Hollandaise sauce (good substitute for mayonnaise but has to be used straight away) *and* béarnaise sauce (*low-carb, keto, Sapiens if the hollandaise and béarnaise sauces are homemade with dripping and a teaspoon of honey*)
- Barbeque sauce, tomato ketchup or brown sauce (in modest amounts) (*low-carb, keto*)
- Mayonnaise (beware that most are made with vegetable oils, so better to choose low-fat/diet versions) (*low-carb, keto*)
- Mint sauce (*low-carb, keto*)

You're all set. *Bon appétit!*

INDEX

Page references in *italics* indicate images.

ABOUT THE AUTHORS

Anita Tagore is a medical doctor (University of Cambridge) with a Master's degree in nutrition (Oxford Brookes University); **Alison Morgan** is a farming and environment specialist with degrees in Agriculture and Global Development (University of Reading); **David Ellis** is a geophysicist with a degree in Natural Sciences (University of Cambridge).

The authors gratefully acknowledge the many friends, family and colleagues who have supported, reviewed and informed the writing of this book.

• • •

Please go to the following website to access the full set of references, notes and sources for this book: http://whyveganshavesmallerbrains.com

www.ingramcontent.com/pod-product-compliance
Lightning Source LLC
LaVergne TN
LVHW100255210926
848480LV00003B/930
* 9 7 8 1 9 1 7 5 2 3 0 5 9 *